DATE DUE

#47-0108 Peel Off Pressure Sensitive

THE GREEN BOOK
OF MATHEMATICAL PROBLEMS

Kenneth Hardy and Kenneth S. Williams
Carleton University, Ottawa

DOVER PUBLICATIONS, INC.
Mineola, New York

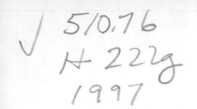

Published in Canada by General Publishing Company, Ltd., 30 Lesmill Road, Don Mills, Toronto, Ontario.
Published in the United Kingdom by Constable and Company, Ltd., 3 The Lanchesters, 162–164 Fulham Palace Road, London W6 9ER.

Bibliographical Note

This Dover edition, first published in 1997, is an unabridged and slightly corrected republication of the work first published by Integer Press, Ottawa, Ontario, Canada in 1985, under the title *The Green Book: 100 Practice Problems for Undergraduate Mathematics Competitions*.

Library of Congress Cataloging-in-Publication Data

Hardy, Kenneth.
    [Green book]
    The green book of mathematical problems / Kenneth Hardy and Kenneth S. Williams.
        p.   cm.
    Originally published: The green book. Ottawa, Ont., Canada : Integer Press, 1985.
    Includes bibliographical references.
    ISBN 0-486-69573-5 (pbk.)
    1. Mathematics—Problems, exercises, etc. I. Williams, Kenneth S. II. Title.
QA43.H268   1997
510'.76—dc21   96-47817
                                                  CIP

Manufactured in the United States of America
Dover Publications, Inc., 31 East 2nd Street, Mineola, N.Y. 11501

PREFACE

There is a famous set of fairy tale books, each volume of which is designated by the colour of its cover: *The Red Book, The Blue Book, The Yellow Book,* etc. We are not presenting you with *The Green Book* of fairy stories, but rather a book of mathematical problems. However, the conceptual idea of all fairy stories, that of mystery, search, and discovery is also found in our *Green Book*. It got its title simply because in its infancy it was contained and grew between two ordinary green file covers.

The book contains 100 problems for undergraduate students training for mathematics competitions, particularly the William Lowell Putnam Mathematical Competition. Along with the problems come useful hints, and in the end (just like in the fairy tales) the solutions to the problems. Although the book is written especially for students training for competitions, it will also be useful to anyone interested in the posing and solving of challenging mathematical problems at the undergraduate level.

Many of the problems were suggested by ideas originating in articles and problems in mathematical journals such as *Crux Mathematicorum, Mathematics Magazine,* and the *American Mathematical Monthly,* as well as problems from the Putnam competition itself. Where possible, acknowledgement to known sources is given at the end of the book.

We would, of course, be interested in your reaction to *The Green Book,* and invite comments, alternate solutions, and even corrections. We make no claims that our solutions are the "best possible" solutions, but we trust you will find them elegant enough, and that *The Green Book* will be a practical tool in the training of young competitors.

We wish to thank our publisher, Integer Press; our literary adviser; and our typist, David Conibear, for their invaluable assistance in this project.

Kenneth Hardy and Kenneth S. Williams
Ottawa, Canada
May, 1985

Dedicated to the contestants of the
William Lowell Putnam Mathematical Competition

**To Carole with love
KSW**

CONTENTS

NOTATION

[x]   denotes the greatest integer $\leq x$, where x is a real number.

{x}   denotes the fractional part of the real number x, that is, $\{x\} = x - [x]$.

ln x   denotes the natural logarithm of x.

exp x   denotes the exponential function of x.

$\varphi(n)$ denotes Euler's totient function defined for any natural number n.

GCD(a,b) denotes the greatest common divisor of the integers a and b.

$\binom{n}{k}$ denotes the binomial coefficient $n!/k!(n-k)!$, where n and k are non-negative integers (the symbol having value zero when $n < k$).

(a_{ij}) denotes a matrix with a_{ij} as the (i,j)th entry.

det A   denotes the determinant of a square matrix A.

THE PROBLEMS

Problems, problems,
problems all day long.
Will my problems work out right or wrong?

The Everly Brothers

1. If $\{b_n : n = 0,1,2,\ldots\}$ is a sequence of non-negative real numbers, prove that the series

$$(1.0) \qquad \sum_{n=0}^{\infty} \frac{b_n}{(a+b_0+b_1+\ldots+b_n)^{3/2}}$$

converges for every positive real number a.

2. Let a,b,c,d be positive real numbers, and let

$$Q_n(a,b,c,d) = \frac{a(a+b)(a+2b)\ldots(a+(n-1)b)}{c(c+d)(c+2d)\ldots(c+(n-1)d)} .$$

Evaluate the limit $L = \lim\limits_{n \to \infty} Q_n(a,b,c,d)$.

3. Prove the following inequality:

$$(3.0) \qquad \frac{\ln x}{x^3-1} < \frac{1}{3} \frac{(x+1)}{(x^3+x)} \quad , \quad x > 0, \; x \neq 1.$$

4. Do there exist non-constant polynomials $p(z)$ in the complex variable z such that $|p(z)| < R^n$ on $|z| = R$, where $R > 0$ and $p(z)$ is monic and of degree n?

5. Let $f(x)$ be a continuous function on $[0,a]$, where $a > 0$, such that $f(x) + f(a-x)$ does not vanish on $[0,a]$. Evaluate the integral

$$\int_0^a \frac{f(x)}{f(x) + f(a-x)} \, dx \ .$$

6. For $\varepsilon > 0$ evaluate the limit

$$\lim_{x \to \infty} x^{1-\varepsilon} \int_x^{x+1} \sin(t^2)\,dt \ .$$

7. Prove that the equation

(7.0) $x^4 + y^4 + z^4 - 2y^2z^2 - 2z^2x^2 - 2x^2y^2 = 24$

has no solution in integers x,y,z.

8. Let a and k be positive numbers such that $a^2 > 2k$. Set $x_0 = a$ and define x_n recursively by

(8.0) $x_n = x_{n-1} + \dfrac{k}{x_{n-1}}$, $n = 1,2,3,\ldots$.

Prove that

$$\lim_{n \to \infty} \frac{x_n}{\sqrt{n}}$$

exists and determine its value.

9. Let x_0 denote a fixed non-negative number, and let a and b be positive numbers satisfying

$$\sqrt{b} < a < 2\sqrt{b} .$$

Define x_n recursively by

(9.0)
$$x_n = \frac{ax_{n-1} + b}{x_{n-1} + a} , \quad n = 1,2,3,\ldots .$$

Prove that $\lim_{n \to \infty} x_n$ exists and determine its value.

10. Let a,b,c be real numbers satisfying

$$a > 0, \ c > 0, \ b^2 > ac .$$

Evaluate

$$\max_{\substack{x,y \,\varepsilon\, R \\ x^2+y^2 = 1}} (ax^2 + 2bxy + cy^2) .$$

11. Evaluate the sum

(11.0)
$$S = \sum_{r=0}^{[n/2]} \frac{n(n-1)\ldots(n-(2r-1))}{(r!)^2} 2^{n-2r}$$

for n a positive integer.

12. Prove that for $m = 0,1,2,\ldots$

(12.0)
$$S_m(n) = 1^{2m+1} + 2^{2m+1} + \ldots + n^{2m+1}$$

is a polynomial in $n(n+1)$.

13. Let a,b,c be positive integers such that

$$GCD(a,b) = GCD(b,c) = GCD(c,a) = 1 .$$

Show that $\ell = 2abc - (bc+ca+ab)$ is the largest integer such that

$$bc\,x + ca\,y + ab\,z = \ell$$

is insolvable in non-negative integers x,y,z.

14. Determine a function $f(n)$ such that the nth term of the sequence

(14.0)   1, 2, 2, 3, 3, 3, 4, 4, 4, 4, 5, ...

is given by $[f(n)]$.

15. Let a_1, a_2, ..., a_n be given real numbers, which are not all zero. Determine the least value of

$$x_1^2 + ... + x_n^2 ,$$

where x_1, ..., x_n are real numbers satisfying

$$a_1x_1 + ... + a_nx_n = 1 .$$

16. Evaluate the infinite series

$$S = 1 - \frac{2^3}{1!} + \frac{3^3}{2!} - \frac{4^3}{3!} + ...$$

17. $F(x)$ is a differentiable function such that $F'(a-x) = F'(x)$ for all x satisfying $0 \leq x \leq a$. Evaluate $\int_0^a F(x)dx$ and give an example of such a function $F(x)$.

18. (a) Let r,s,t,u be the roots of the quartic equation
$$x^4 + Ax^3 + Bx^2 + Cx + D = 0 \ .$$

Prove that if $rs = tu$ then $A^2 D = C^2$.

(b) Let a,b,c,d be the roots of the quartic equation
$$y^4 + py^2 + qy + r = 0 \ .$$

Use (a) to determine the cubic equation (in terms of p,q,r) whose roots are

$$\frac{ab - cd}{a + b - c - d} \ , \quad \frac{ac - bd}{a + c - b - d} \ , \quad \frac{ad - bc}{a + d - b - c} \ .$$

19. Let $p(x)$ be a monic polynomial of degree $m \geq 1$, and set
$$f_n(x) = e^{p(x)} D^n (e^{-p(x)}) \ ,$$

where n is a non-negative integer and $D \equiv \frac{d}{dx}$ denotes differentiation with respect to x.

   Prove that $f_n(x)$ is a polynomial in x of degree $(mn - n)$. Determine the ratio of the coefficient of x^{mn-n} in $f_n(x)$ to the constant term in $f_n(x)$.

20. Determine the real function of x whose power series is
$$\frac{x^3}{3!} + \frac{x^9}{9!} + \frac{x^{15}}{15!} + \cdots \ .$$

21. Determine the value of the integral

(21.0)
$$I_n = \int_0^{\pi} \left(\frac{\sin nx}{\sin x} \right)^2 dx \ ,$$

for all positive integral values of n .

22. During the year 1985, a convenience store, which was open 7 days a week, sold at least one book each day, and a total of 600 books over the entire year. Must there have been a period of consecutive days when exactly 129 books were sold?

23. Find a polynomial $f(x,y)$ with rational coefficients such that as m and n run through all positive integral values, $f(m,n)$ takes on all positive integral values once and once only.

24. Let m be a positive squarefree integer. Let R,S be positive integers. Give a condition involving R,S,m which guarantees that there do not exist rational numbers x,y,z and w such that

$$(24.0) \qquad R + 2S\sqrt{m} \; = \; (x + y\sqrt{m})^2 + (z + w\sqrt{m})^2 \; .$$

25. Let k and h be integers with $1 \leq k < h$. Evaluate the limit

$$(25.0) \qquad L = \lim_{n \to \infty} \prod_{r=kn+1}^{hn} \left(1 - \frac{r}{n^2} \right) \; .$$

26. Let $f(x)$ be a continuous function on $[0,a]$ such that $f(x)f(a-x) = 1$, where $a > 0$. Prove that there exist infinitely many such functions $f(x)$, and evaluate

$$\int_0^a \frac{dx}{1 + f(x)} \; .$$

27. The positive numbers $a_1, a_2, a_3, \ldots$ satisfy

$$(27.0) \qquad \sum_{r=1}^{n} a_r^3 = \left(\sum_{r=1}^{n} a_r \right)^2 \; , \qquad n = 1,2,3,\ldots \; .$$

Is it true that $a_r = r$ for $r = 1,2,3,\ldots$?

28. Let $p > 0$ be a real number and let n be a non-negative integer. Evaluate

(28.0) $$u_n(p) = \int_0^\infty e^{-px} \sin^n x \, dx .$$

29. Evaluate

(29.0) $$\sum_{r=0}^{n-2} 2^r \tan \frac{\pi}{2^{n-r}} ,$$

for integers $n \geq 2$.

30. Let $n \geq 2$ be an integer. A selection $\{s = a_i : i=1,2,\ldots,k\}$ of k ($2 \leq k \leq n$) elements from the set $N = \{1,2,3,\ldots,n\}$ such that $a_1 < a_2 < \cdots < a_k$ is called a k-selection. For any k-selection S , define

$$W(S) = \min \{a_{i+1} - a_i : i = 1,2,\ldots,k-1\} .$$

If a k-selection S is chosen at random from N , what is the probability that

$$W(S) = r ,$$

where r is a natural number?

31. Let $k \geq 2$ be a fixed integer. For $n = 1,2,3,\ldots$ define

$$a_n = \begin{cases} 1 , & \text{if } n \text{ is not a multiple of } k , \\ -(k-1) , & \text{if } n \text{ is a multiple of } k . \end{cases}$$

Evaluate the series $$\sum_{n=1}^\infty \frac{a_n}{n} .$$

32. Prove that

$$\int_0^\infty x^m e^{-x} \sin x \, dx = \frac{m!}{2^{(m+2)/2}} \sin \, (m+1)\pi/4$$

for $m = 0,1,2,\ldots$.

33. For a real number u set

(33.0) $I(u) = \int_0^\pi \ell n(1 - 2u \cos x + u^2) \, dx$.

Prove that

$$I(u) = I(-u) = \frac{1}{2}I(u^2) \, ,$$

and hence evaluate $I(u)$ for all values of u .

34. For each natural number $k \geq 2$ the set of natural numbers is partitioned into a sequence of sets $\{A_n(k): n = 1,2,3,\ldots \}$ as follows: $A_1(k)$ consists of the first k natural numbers, $A_2(k)$ consists of the next $k+1$ natural numbers, $A_3(k)$ consists of the next $k+2$ natural numbers, etc. The sum of the natural numbers in $A_n(k)$ is denoted by $s_n(k)$. Determine the least value of $n = n(k)$ such that $s_n(k) > 3k^3 - 5k^2$, for $k = 2,3,\ldots$.

35. Let $\{p_n: n = 1,2,3,\ldots \}$ be a sequence of real numbers such that $p_n \geq 1$ for $n = 1,2,3,\ldots$. Does the series

(35.0) $\sum_{n=1}^\infty \frac{[p_n]-1}{([p_1]+1)([p_2]+1)\ldots([p_n]+1)}$

converge?

36. Let $f(x)$, $g(x)$ be polynomials with real coefficients of degrees $n+1$, n respectively, where $n \geq 0$, and with positive

leading coefficients A , B respectively. Evaluate

$$L = \lim_{x \to \infty} g(x) \int_0^x e^{f(t)-f(x)} dt \ ,$$

in terms of A, B and n .

37. The lengths of two altitudes of a triangle are h and k , where h ≠ k . Determine upper and lower bounds for the length of the third altitude in terms of h and k .

38. Prove that

$$P_{n,r} = P_{n,r}(x) = \frac{(1-x^{n+1})(1-x^{n+2})\ldots(1-x^{n+r})}{(1-x)(1-x^2)\ldots(1-x^r)}$$

is a polynomial in x of degree nr , where n and r are non-negative integers. (When r = 0 the empty product is understood to be 1 and we have $P_{n,0} = 1$ for all n ≥ 0 .)

39. Let A, B, C, D, E be integers such that B ≠ 0 and

$$F = AD^2 - BCD + B^2E \neq 0 \ .$$

Prove that the number N of pairs of integers (x,y) such that

(39.0) $Ax^2 + Bxy + Cx + Dy + E = 0$,

satisfies

$$N \leq 2d(|F|) \ ,$$

where, for integers n ≥ 1 , d(n) denotes the number of positive divisors of n .

40. Evaluate $\displaystyle\sum_{k=1}^{n} \frac{k}{k^4 + k^2 + 1}$.

41. Let $P_m = P_m(n)$ denote the sum of all possible products of m different integers chosen from the set $\{1,2,\ldots,n\}$. Find formulae for $P_2(n)$ and $P_3(n)$.

42. For $a > b > 0$, evaluate the integral

(42.0)
$$\int_0^\infty \frac{e^{ax} - e^{bx}}{x(e^{ax}+1)(e^{bx}+1)} \, dx \ .$$

43. For integers $n \geq 1$, determine the sum of n terms of the series

(43.0)
$$\frac{2n}{2n-1} + \frac{2n(2n-2)}{(2n-1)(2n-3)} + \frac{2n(2n-2)(2n-4)}{(2n-1)(2n-3)(2n-5)} + \cdots \ .$$

44. Let m be a fixed positive integer and let $z_1, z_2, \ldots, z_k$ be k (≥ 1) complex numbers such that

(44.0)
$$z_1^s + z_2^s + \ldots + z_k^s = 0 \ ,$$

for all $s = m, m+1, m+2, \ldots, m+k-1$. Must $z_i = 0$ for $i = 1,2,\ldots,k$?

45. Let $A_n = (a_{ij})$ be the $n \times n$ matrix where

$$a_{ij} = \begin{cases} x \ , & \text{if } i = j \ , \\ 1 \ , & \text{if } |i-j| = 1 \ , \\ 0 \ , & \text{otherwise,} \end{cases}$$

where $x > 2$. Evaluate $D_n = \det A_n$.

46. Determine a necessary and sufficient condition for the equations

$$
(46.0) \qquad \left\{ \begin{array}{l} x + y + z = A , \\ x^2 + y^2 + z^2 = B , \\ x^3 + y^3 + z^3 = C , \end{array} \right.
$$

to have a solution with at least one of x, y, z equal to zero.

47. Let S be a set of k distinct integers chosen from $1, 2, 3, \ldots, 10^n - 1$, where n is a positive integer. Prove that if

$$
(47.0) \qquad n < \ln\left[\frac{(2^k - 1)}{k} + \frac{(k+1)}{2}\right] \Big/ \ln 10 ,
$$

it is possible to find 2 disjoint subsets of S whose members have the same sum.

48. Let n be a positive integer. Is it possible for $6n$ distinct straight lines in the Euclidean plane to be situated so as to have at least $6n^2 - 3n$ points where exactly three of these lines intersect and at least $6n+1$ points where exactly two of these lines intersect?

49. Let S be a set with n (≥ 1) elements. Determine an explicit formula for the number $A(n)$ of subsets of S whose cardinality is a multiple of 3 .

50. For each integer $n \geq 1$, prove that there is a polynomial $P_n(x)$ with integral coefficients such that

$$
x^{4n}(1-x)^{4n} = (1+x^2)P_n(x) + (-1)^n 4^n .
$$

Define the rational number a_n by

$$
(50.0) \qquad a_n = \frac{(-1)^{n-1}}{4^{n-1}} \int_0^1 P_n(x) \, dx , \qquad n = 1, 2, \ldots .
$$

Prove that a_n satisfies the inequality

$$\left| \pi - a_n \right| < \frac{1}{4^{5n-1}} \, , \quad n = 1,2,\ldots \quad .$$

51. In last year's boxing contest, each of the 23 boxers from the blue team fought exactly one of the 23 boxers from the green team, in accordance with the contest regulation that opponents may only fight if the absolute difference of their weights is less than one kilogram.

Assuming that this year the members of both teams remain the same as last year and that their weights are unchanged, show that the contest regulation is satisfied if the lightest member of the blue team fights the lightest member of the green team, the next lightest member of the blue team fights the next lightest member of the green team, and so on.

52. Let S be the set of all composite positive odd integers less than 79 .

(a) Show that S may be written as the union of three (not necessarily disjoint) arithmetic progressions.

(b) Show that S cannot be written as the union of two arithmetic progressions.

53. For $b > 0$, prove that

$$\left| \int_0^b \frac{\sin x}{x} \, dx - \frac{\pi}{2} \right| < \frac{1}{b} \, ,$$

by first showing that

$$\int_0^b \frac{\sin x}{x} \, dx = \int_0^\infty \left(\int_0^b e^{-ux} \sin x \, dx \right) du \quad .$$

54. Let $a_1, a_2, \ldots, a_{44}$ be 44 natural numbers such that

$$0 < a_1 < a_2 < \ldots < a_{44} \leq 125 .$$

Prove that at least one of the 43 differences $d_j = a_{j+1} - a_j$ occurs at least 10 times.

55. Show that for every natural number n there exists a prime p such that $p = a^2 + b^2$, where a and b are natural numbers both greater than n. (You may appeal to the following two theorems:

(A) If p is a prime of the form $4t+1$ then there exist integers a and b such that $p = a^2 + b^2$.

(B) If r and s are natural numbers such that $GCD(r,s) = 1$, there exist infinitely many primes of the form $rk+s$, where k is a natural number.)

56. Let $a_1, a_2, \ldots, a_n$ be n (≥ 1) integers such that
(1) $0 < a_1 < a_2 < \ldots < a_n$,
(2) all the differences $a_i - a_j$ $(1 \leq j < i \leq n)$ are distinct,
(3) $a_i \equiv a \pmod{b}$ $(1 < i \leq n)$, where a and b are positive integers such that $1 \leq a \leq b-1$.
Prove that

$$\sum_{r=1}^{n} a_r \geq \frac{b}{6} n^3 + \left(a - \frac{b}{6}\right) n .$$

57. Let $A_n = (a_{ij})$ be the $n \times n$ matrix where

$$a_{ij} = \begin{cases} 2 \cos t , & \text{if } i = j , \\ 1 , & \text{if } |i - j| = 1 , \\ 0 , & \text{otherwise} , \end{cases}$$

where $-\pi < t < \pi$. Evaluate $D_n = \det A_n$.

58. Let a and b be fixed positive integers. Find the general solution of the recurrence relation

(58.0) $x_{n+1} = x_n + a + \sqrt{b^2 + 4ax_n}$, $n = 0,1,2,\ldots$,

where $x_0 = 0$.

59. Let a be a fixed real number satisfying $0 < a < \pi$, and set

(59.0) $I_r = \displaystyle\int_{-a}^{a} \frac{1 - r \cos u}{1 - 2r \cos u + r^2} \, du$.

Prove that

$$I_1 \, , \quad \lim_{r \to 1^+} I_r \, , \quad \lim_{r \to 1^-} I_r$$

all exist and are all distinct.

60. Let I denote the class of all isosceles triangles. For $\Delta \in I$, let h_Δ denote the length of each of the two equal altitudes of Δ and k_Δ the length of the third altitude. Prove that there does not exist a function f of h_Δ such that

$$k_\Delta \leq f(h_\Delta) \, ,$$

for all $\Delta \in I$.

61. Find the minimum value of the expression

(61.0) $\left(x^2 + \dfrac{k^2}{x^2} \right) - 2 \left((1 + \cos t)x + \dfrac{k(1 + \sin t)}{x} \right) + (3 + 2\cos t + 2 \sin t)$,

for $x > 0$ and $0 \leq t \leq 2\pi$, where $k > \dfrac{3}{2} + \sqrt{2}$ is a fixed real number.

62. Let $\varepsilon > 0$. Around every point in the xy-plane with integral co-ordinates draw a circle of radius ε. Prove that every straight line through the origin must intersect an infinity of these circles.

63. Let n be a positive integer. For $k = 0,1,2,\ldots,2n-2$ define

(63.0) $$I_k = \int_0^\infty \frac{x^k}{x^{2n}+x^n+1}\, dx \quad.$$

Prove that $I_k \geq I_{n-1}$, $k = 0,1,2,\ldots,2n-2$.

64. Let D be the region in Euclidean n-space consisting of all n-tuples $(x_1,x_2,\ldots,x_n)$ satisfying

$$x_1 \geq 0 \quad, \quad x_2 \geq 0 \quad, \quad \ldots \quad, \quad x_n \geq 0 \quad, \quad x_1 + x_2 + \ldots + x_n \leq 1 \quad.$$

Evaluate the multiple integral

(64.0) $$\iint\cdots\int_D x_1^{k_1} x_2^{k_2} \ldots x_n^{k_n} (1-x_1-x_2-\ldots-x_n)^{k_{n+1}} dx_1 \ldots dx_n \quad,$$

where $k_1,\ldots,k_{n+1}$ are positive integers.

65. Evaluate the limit

$$L = \lim_{n \to \infty} \frac{1}{n} \sum_{k=1}^{n} \left(\left[\frac{2\sqrt{n}}{\sqrt{k}} \right] - 2 \left[\frac{\sqrt{n}}{\sqrt{k}} \right] \right) \quad.$$

66. Let p and q be distinct primes. Let S be the sequence consisting of the members of the set

$$\{p^m q^n: m,n = 0,1,2,\ldots \}$$

arranged in increasing order. For any pair (a,b) of non-negative

integers, give an explicit expression involving a, b, p and q for
the position of $p^a q^b$ in the sequence S .

67. Let p denote an odd prime and let Z_p denote the finite
field consisting of the p elements $0,1,2,\ldots,p-1$. For a an
element of Z_p , determine the number N(a) of 2×2 matrices X ,
with entries from Z_p , such that

(67.0) $X^2 = A$, where $A = \begin{bmatrix} a & 0 \\ 0 & a \end{bmatrix}$.

68. Let n be a non-negative integer and let f(x) be the
unique differentiable function defined for all real x by

(68.0) $(f(x))^{2n+1} + f(x) - x = 0$.

Evaluate the integral

$$\int_0^x f(t)\, dt ,$$

for $x \geq 0$.

69. Let f(n) denote the number of zeros in the usual decimal
representation of the positive integer n , so that for example,
$f(1009) = 2$. For $a > 0$ and N a positive integer, evaluate the
limit

$$L = \lim_{N \to \infty} \frac{\ln S(N)}{\ln N},$$

where

$$S(N) = \sum_{k=1}^{N} a^{f(k)} .$$

70. Let $n \geq 2$ be an integer and let k be an integer with $2 \leq k \leq n$. Evaluate

$$M = \max_{S} \left(\min_{1 \leq i \leq k-1} (a_{i+1} - a_i) \right) ,$$

where S runs over all selections $S = \{a_1, a_2, \ldots, a_k\}$ from $\{1, 2, \ldots, n\}$ such that $a_1 < a_2 < \ldots < a_k$.

71. Let $az^2 + bz + c$ be a polynomial with complex coefficients such that a and b are nonzero. Prove that the zeros of this polynomial lie in the region

$$(71.0) \qquad\qquad |z| \leq \left| \frac{b}{a} \right| + \left| \frac{c}{b} \right| .$$

72. Determine a monic polynomial $f(x)$ with integral coefficients such that $f(x) \equiv 0 \pmod{p}$ is solvable for every prime p but $f(x) = 0$ is not solvable with x an integer.

73. Let n be a fixed positive integer. Determine

$$M = \max_{\substack{0 \leq x_k \leq 1 \\ k=1,2,\ldots,n}} \sum_{1 \leq i < j \leq n} |x_i - x_j| .$$

74. Let $\{x_i : i = 1, 2, \ldots, n\}$ and $\{y_i : i = 1, 2, \ldots, n\}$ be two sequences of real numbers with

$$x_1 \geq x_2 \geq \ldots \geq x_n .$$

How must $y_1, \ldots, y_n$ be rearranged so that the sum

$$(74.0) \qquad\qquad \sum_{i=1}^{n} (x_i - y_i)^2$$

is as small as possible?

75. Let p be an odd prime and let Z_p denote the finite field consisting of $0,1,2,\ldots,p-1$. Let g be a given function on Z_p with values in Z_p. Determine all functions f on Z_p with values in Z_p, which satisfy the functional equation

(75.0) $f(x) + f(x+1) = g(x)$

for all x in Z_p.

76. Evaluate the double integral

(76.0) $I = \int_0^1 \int_0^1 \frac{dxdy}{1 - xy}$.

77. Let a and b be integers and m an integer > 1. Evaluate

$$\left[\frac{b}{m}\right] + \left[\frac{a+b}{m}\right] + \left[\frac{2a+b}{m}\right] + \ldots + \left[\frac{(m-1)a+b}{m}\right] .$$

78. Let $a_1,\ldots,a_n$ be n (>1) distinct real numbers. Set

$$S = a_1^2 + \ldots + a_n^2 , \qquad M = \min_{1 \le i < j \le n} (a_i - a_j)^2 .$$

Prove that

$$\frac{S}{M} \ge \frac{n(n-1)(n+1)}{12} .$$

79. Let $x_1,\ldots,x_n$ be n real numbers such that

$$\sum_{k=1}^n |x_k| = 1 , \qquad \sum_{k=1}^n x_k = 0 .$$

Prove that

(79.0) $\left| \sum_{k=1}^n \frac{x_k}{k} \right| \le \frac{1}{2} - \frac{1}{2n}$.

80. Prove that the sum of two consecutive odd primes is the product of at least three (possibly repeated) prime factors.

81. Let $f(x)$ be an integrable function on the closed interval $[\pi/2, \pi]$ and suppose that

$$\int_{\pi/2}^{\pi} f(x) \sin kx \, dx = \begin{cases} 0 \, , & 1 \le k \le n-1 \, , \\ 1 \, , & k = n \, . \end{cases}$$

Prove that $|f(x)| \ge \dfrac{1}{\pi \ell n \, 2}$ on a set of positive measure.

82. For $n = 0, 1, 2, \ldots$, let

(82.0) $\qquad s_n = \sqrt[3]{a_n + \sqrt[3]{a_{n-1} + \sqrt[3]{a_{n-2} + \ldots + \sqrt[3]{a_0}}}}$

where $a_n = \dfrac{6n+1}{n+1}$. Show that $\lim\limits_{n \to \infty} s_n$ exists and determine its value.

83. Let $f(x)$ be a non-negative strictly increasing function on the interval $[a,b]$, where $a < b$. Let $A(x)$ denote the area below the curve $y = f(x)$ and above the interval $[a,x]$, where $a \le x \le b$, so that $A(a) = 0$.

Let $F(x)$ be a function such that $F(a) = 0$ and

(83.0) $\qquad (x' - x)f(x) < F(x') - F(x) < (x'-x)f(x')$

for all $a \le x < x' \le b$. Prove that $A(x) = F(x)$ for $a \le x \le b$.

84. Let a and b be two given positive numbers with $a < b$. How should the number r be chosen in the interval $[a,b]$ in order to minimize

(84.0) $\qquad M(r) = \max_{a \le x \le b} \left| \dfrac{r - x}{x} \right|$?

85. Let $\{a_n : n = 1, 2, \ldots \}$ be a sequence of positive real numbers with $\lim\limits_{n \to \infty} a_n = 0$ and satisfying the condition $a_n - a_{n+1} > a_{n+1} - a_{n+2} > 0$. For any $\varepsilon > 0$, let N be a positive integer such that $a_N \leq 2\varepsilon$. Prove that $L = \sum\limits_{k=1}^{\infty} (-1)^{k+1} a_k$ satisfies the inequality

$$(85.0) \qquad \left| L - \sum_{k=1}^{N} (-1)^{k+1} a_k \right| < \varepsilon .$$

86. Determine all positive continuous functions $f(x)$ defined on the interval $[0, \pi]$ for which

$$(86.0) \qquad \int_0^{\pi} f(x) \cos nx \, dx = (-1)^n (2n+1) , \quad n = 0, 1, 2, 3, 4 .$$

87. Let P and P' be points on opposite sides of a non-circular ellipse E such that the tangents to E through P and P' respectively are parallel and such that the tangents and normals to E at P and P' determine a rectangle R of maximum area. Determine the equation of E with respect to a rectangular coordinate system, with origin at the centre of E and whose y-axis is parallel to the longer side of R.

88. If four distinct points lie in the plane such that any three of them can be covered by a disk of unit radius, prove that all four points may be covered by a disk of unit radius.

89. Evaluate the sum

$$S = \sum_{\substack{m=1 \\ m \neq n}}^{\infty} \sum_{n=1}^{\infty} \frac{1}{m^2 - n^2} .$$

90. If n is a positive integer which can be expressed in the form $n = a^2 + b^2 + c^2$, where a,b,c are positive integers, prove that, for each positive integer k , n^{2k} can be expressed in the form $A^2 + B^2 + C^2$, where A,B,C are positive integers.

91. Let G be the group generated by a and b subject to the relations $aba = b^3$ and $b^5 = 1$. Prove that G is abelian.

92. Let $\{a_n : n = 1,2,3,\ldots \}$ be a sequence of real numbers satisfying $0 < a_n < 1$ for all n and such that $\sum_{n=1}^{\infty} a_n$ diverges while $\sum_{n=1}^{\infty} a_n^2$ converges. Let $f(x)$ be a function defined on $[0,1]$ such that $f''(x)$ exists and is bounded on $[0,1]$. If $\sum_{n=1}^{\infty} f(a_n)$ converges, prove that $\sum_{n=1}^{\infty} |f(a_n)|$ also converges.

93. Let a,b,c be real numbers such that the roots of the cubic equation

(93.0) $$x^3 + ax^2 + bx + c = 0$$

are all real. Prove that these roots are bounded above by $(2\sqrt{a^2-3b} - a)/3$.

94. Let $Z_5 = \{0,1,2,3,4\}$ denote the finite field with 5 elements. Let a,b,c,d be elements of Z_5 with $a \neq 0$. Prove that the number N of distinct solutions in Z_5 of the cubic equation

$$f(x) = a + bx + cx^2 + dx^3 = 0$$

is given by $N = 4 - R$, where R denotes the rank of the matrix

$$A = \begin{bmatrix} a & b & c & d \\ b & c & d & a \\ c & d & a & b \\ d & a & b & c \end{bmatrix} \; .$$

95. Prove that

(95.0)
$$S = \sum_{\substack{m,n=1 \\ (m,n)=1}}^{\infty} \frac{1}{(mn)^2}$$

is a rational number.

96. Prove that there does not exist a rational function $f(x)$ with real coefficients such that

(96.0)
$$f\left(\frac{x^2}{x+1}\right) = p(x) \; ,$$

where $p(x)$ is a non-constant polynomial with real coefficients.

97. For n a positive integer, set

$$S(n) = \sum_{k=0}^{n} \frac{1}{\binom{n}{k}} \; .$$

Prove that

$$S(n) = \frac{n+1}{2^{n+1}} \sum_{k=1}^{n+1} \frac{2^k}{k} \; .$$

98. Let $u(x)$ be a non-trivial solution of the differential equation

$$u'' + pu = 0 \; ,$$

defined on the interval $I = [1,\infty)$, where $p = p(x)$ is continuous on I . Prove that u has only finitely many zeros in any interval [a,b], $1 \leq a < b$.

(A zero of u(x) is a point z , $1 \leq z < \infty$, with u(z) = 0).

99. Let P_j (j = 0,1,2,...,n-1) be n (≥2) equally spaced points on a circle of unit radius. Evaluate the sum

$$S(n) = \sum_{0 \leq j < k \leq n-1} |P_j P_k|^2 ,$$

where $|PQ|$ denotes the distance between the points P and Q .

100. Let M be a 3×3 matrix with entries chosen at random from the finite field $Z_2 = \{0,1\}$. What is the probability that M is invertible?

THE HINTS

The little fishes of the sea,
   They sent an answer back to me.

The little fishes' answer was
   "We cannot do it, Sir, because ——,"

Lewis Carroll

1. Define

$$a_n = a + b_0 + b_1 + \ldots + b_n \ , \quad n \geq 0,$$

and prove an inequality of the type

$$\frac{a_n - a_{n-1}}{a_n^{3/2}} \leq c\left(\frac{1}{a_{n-1}^{1/2}} - \frac{1}{a_n^{1/2}}\right) , \quad n \geq 1 \ ,$$

where c is a constant.

2. Consider five cases according as
   (a) $b > d$,
   (b) $b = d$ and $a > c$,
   (c) $b < d$,
   (d) $b = d$ and $a < c$,
   (e) $b = d$ and $a = c$.

In case (a) show that $L = +\infty$ by bounding $Q_n(a,b,c,d)$ from below by a multiple of $(\frac{b}{d})^n$. In case (b) show that $L = +\infty$ by estimating $Q_n(a,b,c,d)$ from below in terms of the harmonic series.

Cases (c) and (d) are easily treated by considering $\dfrac{1}{Q_n(a,b,c,d)}$. The final case (e) is trivial.

3. A straightforward approach to this problem is to show that the function

$$F(x) \;=\; \frac{(x^3-1)(x+1)}{(x^3+x)} - 3\ln x \;, \quad x > 0 \;,$$

suggested by the inequality (3.0), is increasing.

4. Apply Rouché's theorem to the polynomials $f(z) = -z^n$ and $g(z) = p(z)$. *Rouché's theorem states that if $f(z)$ and $g(z)$ are analytic within and on a simple closed contour C and satisfy $|g(z)| < |f(z)|$ on C , where $f(z)$ does not vanish, then $f(z)$ and $f(z) + g(z)$ have the same number of zeros inside C .*

5. Apply the change of variable $x = a - t$ to

$$I \;=\; \int_0^a \frac{f(x)}{f(x) + f(a-x)} \; dx \quad .$$

6. Integrate

$$I \;=\; \int_x^{x+1} \frac{2t \sin (t^2)}{2t} \; dt$$

by parts and obtain an upper bound for $|I|$.

7. Consider (7.0) modulo 16.

8. By squaring (8.0), obtain the lower bound $\sqrt{2kn+a^2}$ for x_n. Using this bound in (8.0) obtain an upper bound for x_n .

9. Assume that the required limit exists, and use (9.0) to det-
ermine its value L . Again use (9.0) to estimate $|x_n - L|$.

10. Either set $x = \cos \theta$, $y = \sin \theta$ and maximize the resulting
function of θ , or express $ax^2 + 2bxy + cy^2$ in the form
$A(x^2 + y^2) - (Bx + Cy)^2$ for appropriate constants A,B,C .

11. Consider the coefficient of x^n in both sides of the identity

$$(1 + x)^{2n} \equiv ((1 + 2x) + x^2)^n .$$

12. Express $(\ell(\ell + 1))^k - ((\ell - 1)\ell)^k$, $(k = 1,2,3,...)$ as a
polynomial·in ℓ , then sum over $\ell = 1, 2, ..., n$ to obtain
$(n(n + 1))^k$ as a linear combination of

$$S_{[k/2]}(n), ..., S_{k-1}(n) .$$

Complete the argument using induction.

13. Prove that the equation

$$bc\,x + ca\,y + ab\,z = 2abc - (bc + ca + ab) + k$$

is solvable in non-negative integers x,y,z for every integer $k \geq 1$.
Then show that the equation with k = 0 is insolvable in non-negative
integers x,y,z .

14. Let u_n be the nth term of the sequence (14.0) and show
that $u_n = k$ for $n = \frac{(k-1)k}{2} + 1 + \ell$, $\ell = 0, 1, 2, ..., k-1$,
and deduce that $k \leq \frac{1}{2}(1 + \sqrt{8n-7}) < k + 1$.

15. Use Cauchy's inequality to prove that

$$\sum_{i=1}^{n} x_i^2 \geq \left(\sum_{i=1}^{n} a_i^2 \right)^{-1} \quad ,$$

and choose the x_i so that equality holds.

16. Express $(n + 1)^3$ in the form

$$An(n - 1)(n - 2) + Bn(n - 1) + Cn + D$$

for suitable constants A, B, C, D .

17. Integrate $F'(a-x) = F'(x)$ twice.

18. For part (b), find the quartic equation whose roots are $a - z, b - z, c - z, d - z,$ and use part (a) to ensure that the product of two of these roots is equal to the product of the other two.

19. Differentiate $f_n(x)$ to obtain the difference-differential equation

$$f_{n+1}(x) = f_n'(x) - p'(x) f_n(x) \quad .$$

20. Consider $\sinh x + \sinh wx + \sinh w^2 x$, where $w = \frac{1}{2}(-1 + \sqrt{-3})$.

21. Show that

$$I_n - I_{n-1} = \int_0^{\pi} \frac{\sin(2n - 1) x}{\sin x} \, dx \, , \quad n \geq 2 \, ,$$

and then use a similar idea to evaluate the integral on the right side.

22. Let a_i (i = 1, 2, ..., 365) denote the number of books sold during the period from the first day to the ith day inclusive. Apply Dirichlet's box principle to

$$a_1, a_2, \ldots, a_{365}, a_1 + 129, a_2 + 129, \ldots, a_{365} + 129 .$$

23. Show that a polynomial of the required type is

$$f(x,y) = \frac{(x + y - 1)(x + y - 2)}{2} + x ,$$

by showing that $f(x,y) = k$, where k is a positive integer, has a unique solution in positive integers x and y which may be expressed in terms of the integers r and m defined by

$$\frac{(r - 1)(r - 2)}{2} < k \le \frac{r(r - 1)}{2} , \quad m = k - \frac{(r - 1)(r - 2)}{2} .$$

24. Consider the complex conjugate of (24.0).

25. Consider

$$\ell n \prod_{r=kn+1}^{hn} \left(1 - \frac{r}{n^2} \right)$$

and use the expansion

$$- \ell n (1 - x) = \sum_{k=1}^{\infty} \frac{x^k}{k} , \quad |x| < 1 .$$

26. For the evaluation, set $x = a - y$ in the integral.

27. Use mathematical induction to prove that $a_r = r$ for all positive integers r .

28. Use integration by parts to establish the recurrence relation

$$u_n = \frac{n(n-1)}{n^2+p^2} u_{n-2} \ , \quad n \geq 2 \ .$$

29. The series may be summed by using the identity

$$\tan A = \cot A - 2 \cot 2A \ .$$

30. Prove that the number of k-selections S from N such that
$W(S) \geq r$, $r = 1,2,3,\ldots$, is

$$\binom{n - (k-1)(r-1)}{k} \ .$$

31. For each $n \geq 1$ define integers q_n and r_n uniquely by
$n = kq_n + r_n$, $0 \leq r_n < k$. Express the nth partial sum s_n of
the series in terms of n and q_n , and determine $\lim\limits_{n \to \infty} s_n$ by appealing to the result

$$\lim_{m \to \infty} (1 + \frac{1}{2} + \ldots + \frac{1}{m} - \ln m) = c \ ,$$

where c denotes Euler's constant.

32. Recognize the given integral as the imaginary part of the
integral $\displaystyle\int_0^\infty x^m e^{(i-1)x} dx$. Evaluate the latter integral using integration by parts.

33. For the evaluation, iterate $I(u) = \frac{1}{2}I(u^2)$ to obtain

$$I(u) = \frac{1}{2^n} I(u^{2^n}) \quad (n = 1,2,3,\ldots)$$

and then let $n \to +\infty$ in the case $0 < u < 1$.

34. Determine an exact expression for $s_n(k)$ and then compare the values of $s_{k-1}(k)$ and $s_k(k)$ with $3k^3 - 5k^2$.

35. Show that (35.0) converges by comparison with $\displaystyle\sum_{n=1}^{\infty} \frac{P_n - 1}{P_1 \cdots P_n}$.

36. Apply l'Hôpital's rule.

37. Relate h, k, ℓ to the lengths of the sides of the triangle, and then use the triangle inequality.

38. Obtain the recurrence relation $P_{n+1, r} = P_{n+1, r-1} + x^r P_{n, r}$, and apply the principle of mathematical induction.

39. Show that $Bx + D$ is a divisor of F .

40. The first few terms of the series are

$$\frac{1}{3} = \frac{1}{2}\left(1 - \frac{1}{3}\right) \ ,$$

$$\frac{2}{21} = \frac{1}{2}\left(\frac{1}{3} - \frac{1}{7}\right) \ ,$$

$$\frac{3}{91} = \frac{1}{2}\left(\frac{1}{7} - \frac{1}{13}\right) \ .$$

41. Use the identity

$$(1-x)(1-2x)\ldots(1-nx) = 1 - P_1 x + P_2 x^2 - P_3 x^3 + \ldots + (-1)^n P_n x^n \ .$$

42. For a suitable constant C , set $f(x) = \dfrac{e^x}{e^x + 1} + C$, and

show that for $t > 0$

$$\int_0^t \frac{(e^{ax} - e^{bx})}{x'(e^{ax}+1)(e^{bx}+1)}\, dx = \int_0^t \frac{f(ax)}{x}\, dx - \int_0^t \frac{f(bx)}{x}\, dx \ .$$

43. Let S_n denote the sum of n terms of (43.0). Calculate the first few values of S_n, conjecture the value of S_n in general, and prove it by mathematical induction.

44. Consider the polynomial whose roots are $z_1, z_2, \ldots, z_k$, and use (44.0) to show that its constant term is zero.

45. Obtain a recurrence relation for D_n by expanding D_n by its first row.

46. Express xyz in terms of A, B and C.

47. Consider the sums of the integers in subsets of S and apply Dirichlet's box principle.

48. Count pairs of lines in the proposed configuration.

49. Show that $A(n) = \sum_{\substack{k=0 \\ k\equiv 0 \ (mod\ 3)}}^{n} \binom{n}{k}$ and evaluate this sum by considering $(1+1)^n + (1+w)^n + (1+w^2)^n$, where w is a complex cube root of unity.

50. To prove the required inequality, replace $p_n(x)$ by $\frac{x^{4n}(1-x)^{4n} - (-1)^n 4^n}{1 + x^2}$ in (50.0), and then use the inequalities $\frac{1}{1+x^2} \leq 1$ and $x(1-x) \leq \frac{1}{4}$ to estimate $\int_0^1 \frac{x^{4n}(1-x)^{4n}}{1 + x^2}\, dx$.

51. Let $B_1, B_2, \ldots, B_{23}$ (resp. $G_1, G_2, \ldots, G_{23}$) be the members of the blue (resp. green) team, ordered with respect to increasing weight. For each r ($1 \leq r \leq 23$) consider last year's opponents of $B_{r+1}, \ldots, B_{23}$ or $G_{r+1}, \ldots, G_{23}$ according as B_r is heavier or lighter than G_r.

52. Each member of S can be written in the form $(2r+1)(2r+2s+1)$, for suitable integers $r \geq 1$ and $s \geq 0$. Use this fact to construct the three arithmetic progressions.

53. For $y > 0$ prove that

$$\int_0^y \left(\int_0^b e^{-ux} \sin x \, dx \right) du = \int_0^b (1-e^{-xy}) \frac{\sin x}{x} dx$$

and then show that

$$\lim_{y \to \infty} \int_0^b (1-e^{-xy}) \frac{\sin x}{x} dx = \int_0^b \frac{\sin x}{x} dx .$$

54. Consider $\displaystyle\sum_{j=1}^{43} d_j$.

55. For any natural number n, construct a prime p of the form

$$p = 4k \prod_{r=1}^{n} (r^2+q)^2 - q ,$$

where k is a natural number and $q > n$ is a prime of the form $4t + 3$, so that $p = a^2 + b^2$, $0 < a < b$. Then, assuming $a \leq n$, obtain a contradiction by considering the factor a^2+q of b^2.

56. For $2 \leq r \leq n$ obtain a lower found for $a_r - a_1$ in terms of b and r by considering the differences $a_i - a_j$, $1 \leq j < i \leq r$.

57. Evaluate D_1, D_2, D_3 and conjecture the value of D_n for all n. Prove your conjecture by using the recurrence relation which may be obtained by expanding D_n by its first row.

58. Prove that

$$x_n = x_{n+1} + a - \sqrt{b^2 + 4ax_{n+1}}$$

and use this to obtain the recurrence relation

$$x_{n+1} - 2x_n + x_{n-1} = 2a .$$

59. For $r > 0$ and $r \neq 1$ show that

$$I_r = a + \frac{1}{2}(1-r^2) \int_{-a}^{a} \frac{du}{1 - 2r\cos u + r^2} ,$$

and evaluate the integral using the transformation $t = \tan u/2$.

60. Construct a class of isosceles triangles whose members have two equal altitudes of fixed length h, while their third altitudes are arbitrarily long.

61. Recognize the expression in (61.0) as the square of the distance between a point on a certain circle and a point on another plane curve.

62. When the line L through the origin has irrational slope, use Hurwitz's theorem to obtain an infinity of lattice points whose distances from L are suitably small.

   In 1881 Hurwitz proved the following basic result: If b is an irrational number then there exist infinitely many pairs of integers (m, n) with $n \neq 0$ and GCD(m, n) = 1 such that

$$\left| b - \frac{m}{n} \right| < \frac{1}{\sqrt{5}n^2} \ .$$

This inequality is best possible in the sense that the result becomes false if $\sqrt{5}$ is replaced by any larger constant.

63. Show that $I_k = I_{2n-k-2}$ and use the arithmetic-geometric mean inequality.

64. Express the multiple integral (64.0) as a repeated integral and use the value of $\int_0^a x^r (a-x)^s dx$, where r and s are positive integers and a is a positive real number, successively in the repeated integral.

65. Show that for a suitable integer $f(n)$

$$\sum_{k=1}^{n} \left(\left[\frac{2\sqrt{n}}{\sqrt{k}} \right] - 2 \left[\frac{\sqrt{n}}{\sqrt{k}} \right] \right) = \sum_{s=1}^{f(n)} \left(\left[\frac{4n}{(2s+1)^2} \right] - \left[\frac{4n}{(2s+2)^2} \right] \right) ,$$

and thus compute L in terms of well-known series.

66. $p^a q^b$ is the nth term of the sequence S , where n is the number of pairs of integers (r,s) such that $p^r q^s \le p^a q^b$, $r \ge 0$, $s \ge 0$.

67. A straightforward approach is to determine explicitly all matrices X such that $X^2 = A$. The form of X depends on whether or not a is a square in Z_p .

68. Recall that if $y = g(x)$ is differentiable with positive derivative for $x \ge 0$ and $g(0) = 0$, then

$$\int_0^x g(t)\, dt + \int_0^{g(x)} g^{-1}(t)\, dt = x\, g(x) , \quad x \ge 0 .$$

69. Evaluate $S(10^m-1)$ exactly and use it to estimate $S(N)$.

70. Show that $M = \left[\dfrac{n-1}{k-1}\right]$.

71. Express the roots of $az^2 + bz + c$ in terms of a, b and c and estimate the moduli of these roots.

72. Choose integers a, b and c such that $x^2 + a \equiv 0 \pmod{p}$ is solvable for primes $p \equiv 1 \pmod 4$ and $p = 2$; $x^2 + b \equiv 0 \pmod{p}$ is solvable for $p \equiv 3 \pmod 8$; $x^2 + c \equiv 0 \pmod{p}$ is solvable for $p \equiv 7 \pmod 8$; and set

$$f(x) = (x^2+a)(x^2+b)(x^2+c) .$$

73. Assume without loss of generality that $0 \leq x_1 \leq x_2 \leq \ldots \leq x_n \leq 1$ and show that

$$S = \sum_{1 \leq i < j \leq n} |x_i - x_j| = \sum_{k=1}^{n} x_k (2k-n-1) .$$

Consider those terms in the sum for which $k \geq \frac{1}{2}(n+1)$ and deduce that $M = [n^2/4]$.

74. Show that the smallest sum (74.0) is obtained when the y_i are arranged in decreasing order.

75. Replace x by $x+k$ $(k = 0,1,2,\ldots,p-1)$ in (75.0) and form the alternating sum

$$\sum_{k=0}^{p-1} (-1)^k g(x+k) .$$

76. Express the improper double integral I as a limit of proper double integrals over appropriate subregions of the unit square and use standard methods to show that $I = \pi^2/6$.

77. Use the identity

$$\sum_{x=0}^{k-1}\left[\frac{x}{k} + e\right] = [ek] ,$$

where k is any positive integer and e is any real number.

78. Reorder the a's in ascending order and define $\min\limits_{1\le i\le n} a_i^2 = a_j^2$, for a fixed subscript j .

Set $b_i = a_j + \sqrt{M}(i - j)$ $(i = 1,2,\ldots,n)$ and prove that $a_i^2 \ge b_i^2$. Deduce the required inequality from $S \ge \sum\limits_{i=1}^{n} b_i^2$.

79. Establish and use the inequality

$$\left|\frac{2}{k} - 1 - \frac{1}{n}\right| \le 1 - \frac{1}{n} , \quad 1 \le k \le n .$$

80. Denote the nth prime by p_n , and show that if $p_n + p_{n+1} = 2^k p^\ell$, for some odd prime p , then $k + \ell \ge 3$.

81. Estimate the integral

$$\int_{\pi/2}^{\pi} |f(x)| \left|\sum_{k=1}^{n} \sin kx\right| dx$$

from above under the assumption that $|f(x)| < \frac{1}{\pi \ell n 2}$ on $[\frac{\pi}{2},\pi]$ except for a set of measure 0. Use (81.0) to obtain a lower bound and derive a contradiction.

82. Show that s_n is non-decreasing and bounded above.

83. Assume that $A(x)$ and $F(x)$ differ at some point c in $(a,b]$ and obtain a contradiction by partitioning $[a,c]$ and using (83.0) on each subinterval.

84. A direct approach recognizes $M(r)$ as $\max(\frac{r}{a}-1, 1-\frac{r}{b})$ and then minimizes $M(r)$ with an appropriate choice of r.

85. Let $S_n = \sum_{k=1}^{n} (-1)^{k+1} a_k$ and show that $|S_n - L| < |S_{n-1} - L|$ and $a_n = |S_n - L| + |S_{n-1} - L|$.

86. Express $(\cos^2 x + \cos x + 1)^2$ as a linear combination of $\cos nx$ $(n = 0,1,2,3,4)$ and consider

$$\int_0^\pi f(x)(\cos^2 x + \cos x + 1)^2 \, dx \ .$$

87. Begin by determining R when the ellipse is in standard position and then rotate the axes through an appropriate angle.

88. Recall *Helly's theorem: Given $n(\geq 4)$ convex regions in the plane such that any three have non-empty intersection, then all n regions have non-empty intersection.*

89. Use partial fractions and the result

$$\lim_{N \to \infty} \left(\sum_{k=1}^{N} \frac{1}{k} - \ell n \, N \right) = c \ ,$$

where   c   is Euler's constant, to evaluate

$$\lim_{N \to \infty} \sum_{\substack{n=1 \\ n \neq m}}^{N} \frac{1}{m^2 - n^2} \ .$$

90. Use the identity

$$(x^2 + y^2 + z^2)^2 = (x^2 + y^2 - z^2)^2 + (2xz)^2 + (2yz)^2 \ .$$

91. Prove that   a   and   b   commute by using the relation $aba = b^3$ in the form $b^{-1}ab = b^2 a^{-1}$ to deduce $ab^4 = b^4 a$.

92. Start by applying the extended mean value theorem to   f   on $[0, a_n]$.

93. Let   p   be the largest root of (93.0). Consider the discriminant of $(x^3 + ax^2 + bx + c)/(x - p)$.

94. Let   B   be the Vandermonde matrix given by $B = \begin{bmatrix} 1 & 1 & 1^2 & 1^3 \\ 1 & 2 & 2^2 & 2^3 \\ 1 & 3 & 3^2 & 3^3 \\ 1 & 4 & 4^2 & 4^3 \end{bmatrix}$,   and consider the rank of   BA .

95. Collect together terms having the same value for $GCD(r,s)$ in $\sum_{r,s=1}^{\infty} \frac{1}{(rs)^2}$.

96. Suppose that such a rational function   f(x)   exists and use the decomposition of its numerator and denominator into linear factors to obtain a contradiction.

97. Sum the identity

$$(n+1)! \left[\frac{2}{(n+1)\binom{n}{k}} - \frac{1}{n\binom{n-1}{k}} \right] = k!(n-k)! - (k+1)!(n-k-1)! \ ,$$

for $k = 0,1,2,\ldots,n-1$.

98. Assume that the set of zeros of $u(x)$ on $[a,b]$, $1 \le a < b$, is infinite. Deduce the existence of an accumulation point c in $[a,b]$ with $u(c) = u'(c) = 0$, and then show that $u(x) \equiv 0$ on $[a,b]$.

99. Take P_j $(j = 0,1,2,\ldots,n-1)$ to be the point $\exp(2\pi ji/n)$ on the unit circle $|z| = 1$ in the complex plane, and express $|P_j P_k|^2$ in terms of $\exp(2\pi(k-j)i/n)$.

100. Let $M = (a_{ij})$ $(1 \le i,j \le 3)$ and with the usual notation let $\det M = a_{11}A_{11} + a_{12}A_{12} + a_{13}A_{13}$. Begin by counting the number of triples (a_{11},a_{12},a_{13}) for which $\det M = 0$, distinguishing two cases according as $(A_{11},A_{12},A_{13}) = (0,0,0)$ or not.

THE SOLUTIONS

It seemed that the next minute
  they would discover a solution.
Yet it was clear to both of them
  that the end was still far, far off,
and that the hardest and most complicated
  part was only just beginning.

Anton Chekhov

1. If $\{b_n : n = 0,1,2,\ldots\}$ is a sequence of non-negative real numbers, prove that the series

(1.0)
$$\sum_{n=0}^{\infty} \frac{b_n}{(a+b_0+b_1+\ldots+b_n)^{3/2}}$$

converges for every positive real number a.

Solution: For $a > 0$ we set

$$a_n = a + b_0 + b_1 + \ldots + b_n , \qquad n \geq 0 ,$$

so that

$$a_n - a_{n-1} = b_n , \qquad n \geq 1 .$$

As $b_n \geq 0$ we have, for $n \geq 1$,

$$a_0 \geq a > 0 \quad \text{and} \quad a_n \geq a_{n-1} > 0 .$$

Now, for $n \geq 1$, we may deduce that

$$
\frac{b_n}{(a+b_0+b_1+\ldots+b_n)^{3/2}} = \frac{a_n - a_{n-1}}{a_n^{3/2}}
$$

$$
= \frac{1}{a_n^{1/2}} - \frac{a_{n-1}}{a_n^{3/2}}
$$

$$
= \frac{a_{n-1}}{a_n^{1/2}} \left(\frac{1}{a_{n-1}} - \frac{1}{a_n} \right)
$$

$$
= \frac{a_{n-1}}{a_n^{1/2}} \left(\frac{1}{a_{n-1}^{1/2}} + \frac{1}{a_n^{1/2}} \right) \left(\frac{1}{a_{n-1}^{1/2}} - \frac{1}{a_n^{1/2}} \right)
$$

$$
= \frac{a_{n-1}^{1/2}}{a_n^{1/2}} \left(1 + \frac{a_{n-1}^{1/2}}{a_n^{1/2}} \right) \left(\frac{1}{a_{n-1}^{1/2}} - \frac{1}{a_n^{1/2}} \right)
$$

$$
\leq 2 \left(\frac{1}{a_{n-1}^{1/2}} - \frac{1}{a_n^{1/2}} \right) .
$$

Hence, for $m \geq 1$, we have

$$
s_m = \sum_{n=0}^{m} \frac{b_n}{(a+b_0+b_1+\ldots+b_n)^{3/2}} \leq \frac{b_0}{a_0^{3/2}} + 2 \sum_{n=1}^{m} \left(\frac{1}{a_{n-1}^{1/2}} - \frac{1}{a_n^{1/2}} \right)
$$

$$
= \frac{b_0}{a_0^{3/2}} + 2 \left(\frac{1}{a_0^{1/2}} - \frac{1}{a_m^{1/2}} \right)
$$

$$
< \frac{b_0}{a_0^{3/2}} + \frac{2}{a_0^{1/2}}
$$

$$
\leq \frac{b_0}{a^{3/2}} + \frac{2}{a^{1/2}} .
$$

As the partial sums s_m of (1.0) are bounded, the infinite series (1.0) converges for every $a > 0$.

$2.$ Let a,b,c,d be positive real numbers, and let

$$Q_n(a,b,c,d) = \frac{a(a+b)(a+2b)\ldots(a+(n-1)b)}{c(c+d)(c+2d)\ldots(c+(n-1)d)} .$$

Evaluate the limit $L = \lim\limits_{n \to \infty} Q_n(a,b,c,d)$.

Solution: Considering the five cases specified in THE HINTS, we show
that

$$L = \begin{cases} +\infty, & \text{in cases (a),(b)}, \\ 0, & \text{in cases (c),(d)}, \\ 1, & \text{in case (e)}. \end{cases}$$

We set $k = \left[\dfrac{c}{d}\right] + 1$, so that $c < kd$.

(a) When $b > d$, as $c + jd < (j + k)d$, we have for $n \geq k$

$$Q_n(a,b,c,d) \geq \frac{a(n-1)!\, b^{n-1}}{k(k+1)\ldots(k+(n-1))\, d^n}$$

$$= \frac{a(k-1)!\, b^{n-1}}{(k+n-1)(k+n-2)\ldots n\, d^n}$$

$$\geq \frac{a(k-1)!}{(k+n-1)^k\, d} \cdot \left(\frac{b}{d}\right)^{n-1},$$

which tends to infinity as n tends to infinity, showing that $L = +\infty$.

(b) When $b = d$ and $a > c$ we have

$$Q_n(a,b,c,d) = \prod_{\ell=0}^{n-1} \left(\frac{a + \ell b}{c + \ell b}\right)$$

$$= \prod_{\ell=0}^{n-1} \left(1 + \frac{a - c}{c + \ell b}\right)$$

$$> \prod_{\ell=0}^{n-1} \left(1 + \frac{a - c}{b(\ell+k)}\right) > \frac{a - c}{b} \sum_{\ell=0}^{n-1} \frac{1}{\ell + k},$$

which tends to infinity as n tends to infinity since the harmonic series diverges, showing that $L = +\infty$.

(c)(d) By considering the reciprocal of $Q_n(a,b,c,d)$, we obtain from the above results $L = 0$, if $b < d$, or if $b = d$ and $a < c$.

(e) Clearly $Q_n(a,b,c,d) = 1$ in this case, so that $L = 1$.

3. Prove the following inequality:

(3.0) $$\frac{\ell n\ x}{x^3-1} < \frac{1}{3}\frac{(x+1)}{(x^3+x)} \quad , \quad x > 0, \ x \neq 1.$$

Solution: For $x > 0$ we define

$$F(x) = \frac{(x^3-1)(x+1)}{(x^3+x)} - 3\ell n\ x \ ,$$

so that

$$F(x) = \frac{x^4+x^3-x-1}{(x^3+x)} - 3\ell n\ x \ .$$

Differentiating $F(x)$ with respect to x , we obtain

$$F'(x) = \frac{(4x^3+3x^2-1)(x^3+x) - (x^4+x^3-x-1)(3x^2+1)}{(x^3+x)^2} - \frac{3}{x} \ ,$$

$$= \frac{x^6+3x^4+4x^3+3x^2+1}{(x^3+x)^2} - \frac{3}{x} \ ,$$

that is

(3.1) $$F'(x) = \frac{x^6-3x^5+3x^4-2x^3+3x^2-3x+1}{(x^3+x)^2} \ .$$

The polynomial $p(x)$ in the numerator on the right in (3.1) has the property that $p(x) = x^6 p(\frac{1}{x})$, and so $x^{-3}p(x)$ can be written as a cubic polynomial in $x + 1/x$.

We have

$$F'(x) = \frac{x^3}{(x^3+x)^2}\left[\left(x+\tfrac{1}{x}\right)^3 - 3\left(x+\tfrac{1}{x}\right)^2 + 4\right].$$

As $X^3-3X+4 = (X+1)(X-2)^2$ we obtain

$$F'(x) = \frac{(x^2+x+1)}{(x^2+1)^2}\left[\left(x+\tfrac{1}{x}\right) - 2\right]^2$$

$$= \frac{\left(\left(x+\tfrac{1}{2}\right)^2 + \tfrac{3}{4}\right)(x-1)^2}{x^2(x^2+1)^2} \quad,$$

so that $F'(x) > 0$ for $x > 0$, while $F'(1) = 0$. Thus $F(x)$ is a strictly increasing function of x for all $x > 0$. Hence in particular we have

$$F(x) > F(1) , \quad \text{for} \quad x > 1 ,$$

and so

(3.2) $$\frac{\ell n\ x}{x^3-1} < \frac{1}{3}\frac{(x+1)}{(x^3+x)} \quad, \quad \text{for} \quad x > 1 .$$

Replacing x by $\frac{1}{x}$ in (3.2), we obtain

(3.3) $$\frac{\ell n\ x}{x^3-1} < \frac{1}{3}\frac{(x+1)}{(x^3+x)} \quad, \quad \text{for} \quad 0 < x < 1 .$$

Inequalities (3.2) and (3.3) give the required inequality.

4. Do there exist non-constant polynomials $p(z)$ in the complex variable z such that $|p(z)| < R^n$ on $|z| = R$, where $R > 0$ and $p(z)$ is monic and of degree n ?

Solution: We show that no such polynomial $p(z)$ exists, for suppose there exists a non-constant polynomial

$$p(z) = z^n + a_{n-1}z^{n-1} + \ldots + a_1z + a_0$$

such that $|p(z)| < R^n$ on $|z| = R$.

Then we have

$$\left| z^n + (a_{n-1} z^{n-1} + \ldots + a_1 z + a_0) \right| < \left| -z^n \right| \quad \text{on} \quad |z| = R \ ,$$

and so, by Rouché's theorem,

$$z^n + (a_{n-1} z^{n-1} + \ldots + a_1 z + a_0) - z^n \quad \text{and} \quad -z^n$$

have the same number of zeros counted with respect to multiplicity inside $|z| = R$, that is, $a_{n-1} z^{n-1} + \ldots + a_1 z + a_0$ has n zeros, which is clearly a contradiction. Hence no such polynomial $p(z)$ exists.

5. Let $f(x)$ be a continuous function on $[0,a]$, where $a > 0$, such that $f(x) + f(a-x)$ does not vanish on $[0,a]$. Evaluate the integral

$$\int_0^a \frac{f(x)}{f(x) + f(a-x)} \ dx \ .$$

Solution: Set

$$I = \int_0^a \frac{f(x)}{f(x) + f(a-x)} \ dx \ , \quad J = \int_0^a \frac{f(a-x)}{f(x) + f(a-x)} \ dx \ .$$

Clearly we have

$$I + J = \int_0^a 1 \ dx = a \ .$$

On the other hand, changing the variable from x to a-x in I , we obtain

$$I = J \ .$$

Hence we have

$$I = J = \frac{1}{2} a \ .$$

6. For $\varepsilon > 0$ evaluate the limit

$$\lim_{x \to \infty} x^{1-\varepsilon} \int_x^{x+1} \sin(t^2)\, dt \ .$$

Solution: Integrating by parts we obtain

$$\int \sin(t^2)\, dt = \frac{-\cos(t^2)}{2t} - \frac{1}{2} \int \frac{\cos(t^2)}{t^2}\, dt$$

so that for $x > 0$

$$\int_x^{x+1} \sin(t^2)\, dt = \frac{-\cos(x+1)^2}{2(x+1)} + \frac{\cos x^2}{2x} - \frac{1}{2} \int_x^{x+1} \frac{\cos(t^2)}{t^2}\, dt$$

giving

$$\left| \int_x^{x+1} \sin(t^2)\, dt \right| \le \frac{1}{2(x+1)} + \frac{1}{2x} + \frac{1}{2} \int_x^{x+1} \frac{dt}{t^2} = \frac{1}{x}$$

so that

$$\left| x^{1-\varepsilon} \int_x^{x+1} \sin(t^2)\, dt \right| \le \frac{1}{x^\varepsilon} \ .$$

Since $\frac{1}{x^\varepsilon} \to 0$, as $x \to +\infty$, we deduce that

$$\lim_{x \to \infty} x^{1-\varepsilon} \int_x^{x+1} \sin(t^2)\, dt = 0 \ .$$

7. Prove that the equation

(7.0) $x^4 + y^4 + z^4 - 2y^2 z^2 - 2z^2 x^2 - 2x^2 y^2 = 24$

has no solutions in integers x, y, z .

Solution: Suppose that (7.0) is solvable in integers x, y and z .
    Clearly $x^4 + y^4 + z^4$ must be even. However x,y,z cannot
all be even, as 24 is not divisible by 16 . Hence exactly one of
x,y,z is even, and, without loss of generality, we may suppose that

$$x \equiv 0 \pmod 2 , \quad y \equiv z \equiv 1 \pmod 2 .$$

Thus we have

$$x^4 \equiv 0 \pmod{16} , \quad y^4 \equiv z^4 \equiv 1 \pmod{16} ,$$

$$-2y^2z^2 \equiv -2 \pmod{16} , \quad -2z^2x^2 \equiv -2x^2y^2 \equiv -2x^2 \pmod{16},$$

and so (7.0) gives

$$-4x^2 \equiv 8 \pmod{16} ,$$

that is

$$x^2 \equiv 2 \pmod 4 ,$$

which is impossible.

Second solution: We begin by expressing the left side of (7.0) as
        the product of four linear factors. It is easy to
check that

$$A^2 + B^2 + C^2 - 2BC - 2CA - 2AB = (A+B-C)^2 - 4AB$$
$$= \left((A+B-C) - 2\sqrt{AB} \right)\left((A+B-C) + 2\sqrt{AB} \right) .$$

Replacing A,B,C by x^2, y^2, z^2 respectively, we obtain

$$x^4 + y^4 + z^4 - 2y^2z^2 - 2z^2x^2 - 2x^2y^2$$
$$= (x^2 + y^2 - z^2 - 2xy)(x^2 + y^2 - z^2 + 2xy)$$
$$= ((x-y)^2 - z^2)((x+y)^2 - z^2)$$
$$= (x - y - z)(x - y + z)(x + y - z)(x + y + z) ,$$

so that (7.0) becomes

$$(x - y - z)(x - y + z)(x + y - z)(x + y + z) = 24 .$$

In view of the form of the left side of (7.0), we may assume without loss of generality that any solution (x,y,z) satisfies $x \geq y \geq z \geq 1$, so that

$$x - y - z \leq x - y + z \leq x + y - z \leq x + y + z .$$

Moreover $x - y - z$ and $x - y + z$ cannot both be 1 . As $24 = 2^3 \cdot 3$, we have

$$(x-y-z, \; x-y+z, \; x+y-z, \; x+y+z)$$
$$= (1,2,2,6), (1,2,3,4) \text{ or } (2,2,2,3).$$

However none of the resulting linear systems is solvable in positive integers x,y,z .

8. Let a and k be positive numbers such that $a^2 > 2k$. Set $x_0 = a$ and define x_n recursively by

$$(8.0) \qquad\qquad x_n = x_{n-1} + \frac{k}{x_{n-1}} , \quad n = 1,2,3,\ldots .$$

Prove that

$$\lim_{n \to \infty} \frac{x_n}{\sqrt{n}}$$

exists and determine its value.

<u>Solution</u>: We will show that

$$\lim_{n \to \infty} \frac{x_n}{\sqrt{n}} = \sqrt{2k} .$$

Clearly $x_n > 0$ for all $n \geq 0$. Since $x_n = x_{n-1} + \frac{k}{x_{n-1}}$ for $n = 1,2,\ldots$, we have

$$(8.1) \qquad\qquad x_n^2 = x_{n-1}^2 + 2k + \frac{k^2}{x_{n-1}^2}$$

and so

$$x_n^2 > x_{n-1}^2 + 2k > x_{n-2}^2 + 4k > x_{n-3}^2 + 6k > \ldots > x_0^2 + 2kn = a^2 + 2kn \ ,$$

that is

(8.2) $x_n \geq \sqrt{2kn + a^2} \ , \quad n = 0,1,2,\ldots \ .$

On the other hand, we have, using (8.1) and (8.2),

$$x_n^2 \leq x_{n-1}^2 + 2k + \frac{k^2}{2k(n-1)+a^2} \ , \quad n = 1,2,\ldots \ ,$$

and thus

$$x_n^2 \leq x_0^2 + 2kn + k^2 \sum_{i=0}^{n-1} \frac{1}{2ki + a^2}$$

$$\leq a^2 + 2kn + k^2 \int_{-1}^{n-1} \frac{dx}{2kx + a^2}$$

$$= 2kn + a^2 + \frac{k}{2} \ln\left(\frac{2k(n-1) + a^2}{a^2 - 2k}\right)$$

giving

(8.3) $x_n \leq \sqrt{2kn + a^2 + \frac{k}{2} \ln\left(\frac{2kn + (a^2 - 2k)}{a^2 - 2k}\right)} \ , \quad n = 0,1,2,\ldots \ .$

Hence, from (8.2) and (8.3), we obtain

$$1 \leq \frac{x_n}{\sqrt{2kn+a^2}} \leq \sqrt{1 + \frac{k}{2} \frac{\ln\left(\frac{2kn + (a^2 - 2k)}{a^2 - 2k}\right)}{2kn + a^2}}$$

and thus

$$\lim_{n \to \infty} \frac{x_n}{\sqrt{2kn + a^2}} = 1 \ .$$

Since $\displaystyle\lim_{n \to \infty} \frac{\sqrt{2kn + a^2}}{\sqrt{n}} = \sqrt{2k}$, we obtain $\displaystyle\lim_{n \to \infty} \frac{x_n}{\sqrt{n}} = \sqrt{2k}$.

9. Let x_0 denote a fixed non-negative number, and let a and b be positive numbers satisfying

$$\sqrt{b} < a < 2\sqrt{b} \ .$$

Define x_n recursively by

(9.0)
$$x_n = \frac{ax_{n-1} + b}{x_{n-1} + a} \ , \quad n = 1,2,3,\ldots \ .$$

Prove that $\lim\limits_{n \to \infty} x_n$ exists and determine its value.

Solution: As $x_0 \geq 0$, $a > 0$, $b > 0$, the recurrence relation shows that $x_n > 0$ for $n = 1,2,\ldots$. If $\lim\limits_{n \to \infty} x_n$ exists, say equal to L , then from (9.0) we obtain

$$L = \frac{aL + b}{L + a} \ ,$$

so that $L^2 = b$, $L = +\sqrt{b}$.

Next we have

$$\begin{aligned}
|x_n - \sqrt{b}| &= \left| \frac{ax_{n-1} + b}{x_{n-1} + a} - \sqrt{b} \right| \\
&= \left| \frac{(a - \sqrt{b})(x_{n-1} - \sqrt{b})}{x_{n-1} + a} \right| \\
&= \frac{(a - \sqrt{b})|x_{n-1} - \sqrt{b}|}{x_{n-1} + a} \\
&\leq \frac{(a - \sqrt{b})}{a}|x_{n-1} - \sqrt{b}| \\
&\leq \frac{|x_{n-1} - \sqrt{b}|}{2} \ ,
\end{aligned}$$

so that

$$|x_n - \sqrt{b}| \leq \frac{|x_0 - \sqrt{b}|}{2^n} \ .$$

Letting n tend to infinity, we obtain

$$\lim_{n \to \infty} x_n = \sqrt{b} \ .$$

10. Let a,b,c be real numbers satisfying

$$a > 0, \ c > 0, \ b^2 > ac \ .$$

Evaluate

$$\max_{\substack{x,y \in R \\ x^2+y^2=1}} (ax^2 + 2bxy + cy^2) \ .$$

Solution: All pairs $(x,y) \in R \times R$ satisfying $x^2 + y^2 = 1$ are given
by $x = \cos \theta$, $y = \sin \theta$, $0 \leq \theta \leq 2\pi$. Hence we have

$$\max_{\substack{(x,y) \in R^2 \\ x^2+y^2=1}} (ax^2 + 2bxy + cy^2) = \max_{0 \leq \theta \leq 2\pi} F(\theta) \ ,$$

where

$$\begin{aligned} F(\theta) &= a \cos^2\theta + 2b \cos \theta \sin \theta + c \sin^2\theta \\ &= \frac{a}{2}(1 + \cos 2\theta) + b \sin 2\theta + \frac{c}{2}(1 - \cos 2\theta) \\ &= \frac{1}{2}(a+c) + b \sin 2\theta + \frac{1}{2}(a-c) \cos 2\theta \\ &= \frac{1}{2}(a+c) + \frac{1}{2}\sqrt{(a-c)^2+4b^2} \ \sin (2\theta + \alpha) \ , \end{aligned}$$

where

$$\tan \alpha = \frac{a - c}{2b} \ .$$

Clearly $\max_{0 \leq \theta \leq 2\pi} F(\theta)$ is attained when $\sin (2\theta + \alpha) = 1$, and the
required maximum is

$$\frac{1}{2}(a+c) + \frac{1}{2}\sqrt{(a-c)^2 + 4b^2} \ .$$

Second solution: We seek real numbers A,B,C such that

$$(10.1) \qquad ax^2 + 2bxy + cy^2 \equiv A(x^2+y^2) - (Bx+Cy)^2 \ .$$

Equating coefficients we obtain

$$(10.2) \qquad\qquad\qquad A - B^2 = a \ ,$$

$$(10.3) \qquad\qquad\qquad -2BC = 2b \ ,$$

$$(10.4) \qquad\qquad\qquad A - C^2 = c \ .$$

Subtracting (10.2) from (10.4) we obtain

(10.5) $B^2 - C^2 = c - a$.

Then, from (10.3) and (10.5), we have

$$(B^2 + C^2)^2 = (B^2 - C^2)^2 + (2BC)^2$$
$$= (c - a)^2 + 4b^2 ,$$

so that

(10.6) $B^2 + C^2 = +\sqrt{(a-c)^2 + 4b^2}$.

Adding and subtracting (10.5) and (10.6), and taking square roots, we get

(10.7) $B = \sqrt{\dfrac{\sqrt{(a-c)^2 + 4b^2} - (a-c)}{2}}$, $C = \sqrt{\dfrac{\sqrt{(a-c)^2 + 4b^2} + (a-c)}{2}}$.

Then, from (10.2) and (10.7), we have

$$A = \frac{1}{2}\left(\sqrt{(a-c)^2 + 4b^2} + (a+c)\right) .$$

Finally, from (10.1), we see that the largest value of $ax^2+2bxy+cy^2$ on the circle $x^2+y^2=1$ occurs when $Bx + Cy = 0$, that is, at the points

$$(x,y) = \left(\frac{\pm C}{B^2 + C^2} , \frac{\mp B}{B^2 + C^2}\right) ,$$

and we have

$$\max_{x^2+y^2=1} (ax^2+2bxy+cy^2) = A = \frac{1}{2}\left(\sqrt{(a-c)^2+4b^2} + (a+c)\right) .$$

11. Evaluate the sum

(11.0)
$$S = \sum_{r=0}^{[n/2]} \frac{n(n-1)\ldots(n-(2r-1))}{(r!)^2} 2^{n-2r}$$

for n a positive integer.

Solution: We have

$$S = \sum_{r=0}^{[n/2]} \frac{n!}{(r!)^2(n-2r)!} 2^{n-2r}$$

$$= \sum_{r=0}^{[n/2]} \binom{n}{n-r}\binom{n-r}{n-2r} 2^{n-2r}$$

$$= \sum_{\frac{n}{2} \le s \le n} \binom{n}{s}\binom{s}{2s-n} 2^{2s-n}$$

$$= \sum_{\substack{s=0 \\ 2s-t=n}}^{n} \sum_{t=0}^{s} \binom{n}{s}\binom{s}{t} 2^t ,$$

which is the coefficient of x^n in

$$F(x) = \sum_{s=0}^{n} \sum_{t=0}^{s} \binom{n}{s}\binom{s}{t} 2^t x^{2n-2s+t} .$$

Now

$$F(x) = \sum_{s=0}^{n} \binom{n}{s}\left\{\sum_{t=0}^{s} \binom{s}{t} 2^t x^t\right\}(x^2)^{n-s}$$

$$= \sum_{s=0}^{n} \binom{n}{s} (1 + 2x)^s (x^2)^{n-s}$$

$$= ((1 + 2x) + x^2)^n$$

$$= (1 + x)^{2n} .$$

As the coefficient of x^n in $(1+x)^{2n}$ is $\binom{2n}{n}$, we have $S = \frac{(2n!)}{(n!)^2}$.

<u>12.</u> Prove that for m = 0,1,2,...

(12.0) $S_m(n) = 1^{2m+1} + 2^{2m+1} + \ldots + n^{2m+1}$

is a polynomial in n(n+1) .

<u>Solution:</u> We prove that

$$2\binom{1}{0}S_0(n) = n(n+1) \quad ,$$

$$2\binom{2}{1}S_1(n) = (n(n+1))^2 \quad ,$$

$$2\binom{3}{0}S_1(n) + 2\binom{3}{2}S_2(n) = (n(n+1))^3 \quad ,$$

$$2\binom{4}{1}S_2(n) + 2\binom{4}{3}S_3(n) = (n(n+1))^4 \quad ,$$

$$\vdots$$

and generally for k = 1,2,3,...

(12.1) $$2\sum_{\substack{r=0 \\ r+k \text{ odd}}}^{k}\binom{k}{r}S_{\frac{(r+k-1)}{2}}(n) = (n(n+1))^k \quad .$$

An easy induction argument then shows that $S_m(n)$ (m = 0,1,2,...) is a polynomial in n(n+1).

We now prove (12.1). We have

$$2\sum_{\substack{r=0 \\ r+k \text{ odd}}}^{k}\binom{k}{r}S_{\frac{r+k-1}{2}}(n) = 2\sum_{\substack{r=0 \\ r+k \text{ odd}}}^{k}\binom{k}{r}\sum_{\ell=1}^{n}\ell^{r+k}$$

$$= 2\sum_{\ell=1}^{n}\sum_{\substack{r=0 \\ r+k \text{ odd}}}^{k}\binom{k}{r}\ell^{r+k}$$

$$= \sum_{\ell=1}^{n} \sum_{r=0}^{k} \binom{k}{r} \ell^{2r} (\ell^{k-r} - (-\ell)^{k-r})$$

$$= \sum_{\ell=1}^{n} ((\ell^2+\ell)^k - (\ell^2-\ell)^k)$$

$$= \sum_{\ell=1}^{n} ((\ell(\ell+1))^k - ((\ell-1)\ell)^k)$$

$$= (n(n+1))^k$$

as required.

13. Let a,b,c be positive integers such that

$$GCD(a,b) = GCD(b,c) = GCD(c,a) = 1 .$$

Show that $\ell = 2abc - (bc+ca+ab)$ is the largest integer such that

$$bc\,x + ca\,y + ab\,z = \ell$$

is insolvable in non-negative integers x,y,z .

Solution: We begin by proving the following simple fact which will
            be needed below:

Let A,B,C, be real numbers such that

$$A + B + C < -2 .$$

Then there exist integers t,u,v satisfying

$$t - u > A ,$$
$$u - v > B ,$$
$$v - t > C .$$

To see this, choose

$$t = [A] + 1 ,$$
$$u = 0 ,$$
$$v = -[B] - 1 ,$$

so that t, u, v are integers with

$$t - u = [A] + 1 > A ,$$
$$u - v = [B] + 1 > B ,$$
$$v - t = -[A] - [B] - 2 \geq -A - B - 2 > C .$$

The required result will follow from the two results below:

(a) If k is an integer ≥ 1, then

$$bc\, x + ca\, y + ab\, z = 2abc - (bc + ca + ab) + k$$

is always solvable in non-negative integers x, y, z .

(b) The equation $bc\, x + ca\, y + ab\, z = 2abc - (bc + ca + ab)$ is insolvable in non-negative integers x, y, z .

Proof of (a): As $GCD(ab, bc, ca) = 1$, there exist integers x_0, y_0, z_0 such that

$$bc\, x_0 + ca\, y_0 + ab\, z_0 = k .$$

Take

$$A = -\frac{x_0}{a} - 1 ,$$

$$B = -\frac{y_0}{b} - 1 ,$$

$$C = -\frac{z_0}{c} ,$$

so that

$$A + B + C = -\left(\frac{x_0}{a} + \frac{y_0}{b} + \frac{z_0}{c}\right) - 2$$

$$= -\frac{k}{abc} - 2$$

$$< -2 .$$

Hence, by our initial simple fact, there are integers t,u,v such that

$$t - u > -\frac{x_0}{a} - 1 ,$$

$$u - v > -\frac{y_0}{b} - 1 ,$$

$$v - t > -\frac{z_0}{c} \qquad .$$

Thus we have

$$a + x_0 + at - au > 0 ,$$
$$b + y_0 + bu - bv > 0 ,$$
$$z_0 + cv - ct > 0 .$$

Set

$$x = a-1 + x_0 + at - au ,$$
$$y = b-1 + y_0 + bu - bv ,$$
$$z = -1 + z_0 + cv - ct ,$$

so that x,y,z are non-negative integers.

   Moreover

$$bc\,x + ca\,y + ab\,z = 2abc - (ab + bc + ca) + k$$

as required.

Proof of (b): Suppose the equation is solvable, then

$$2abc = bc(x+1) + ca(y+1) + ab(z+1) ,$$

where x+1,y+1,z+1 are positive integers.

Clearly, as $GCD(a,b) = GCD(b,c) = GCD(c,a) = 1$, we have that a divides $x+1$, b divides $y+1$, and c divides $z+1$. Thus there are positive integers r,s,t such that

$$x+1 = ar, \quad y+1 = bs, \quad z+1 = ct.$$

Hence we have

$$2abc = abc(r + s + t),$$

that is

$$2 = r + s + t \geq 3,$$

which is impossible.

This completes the solution.

14. Determine a function $f(n)$ such that the nth term of the sequence

$$(14.0) \qquad 1, 2, 2, 3, 3, 3, 4, 4, 4, 4, 5, \ldots$$

is given by $[f(n)]$.

Solution: Let u_n be the nth term of the sequence (14.0). The integer k first occurs in the sequence when

$$n = 1 + 2 + 3 + \ldots + (k-1) + 1 = \frac{(k-1)k}{2} + 1.$$

Hence $u_n = k$ for

$$(14.1) \qquad n = \frac{(k-1)k}{2} + 1 + \ell, \quad \ell = 0,1,2,\ldots,k-1.$$

From (14.1) we obtain

$$0 \leq n - \frac{(k-1)k}{2} - 1 \leq k-1$$

and so

$$(14.2) \qquad \frac{k^2 - k + 2}{2} \leq n \leq \frac{k^2 + k}{2}.$$

Multiplying (14.2) by 8 and completing the square, we have

$$(2k-1)^2 + 7 \leq 8n \leq (2k+1)^2 - 1 \; ,$$

$$(2k-1)^2 \leq 8n - 7 \leq (2k+1)^2 - 8 < (2k+1)^2 \; ,$$

$$2k-1 \leq \sqrt{8n-7} < 2k+1 \; ,$$

$$k \leq \frac{1 + \sqrt{8n-7}}{2} < k + 1 \; ;$$

that is

$$u_n = k = [(1 + \sqrt{8n-7})/2] \; .$$

15. Let $a_1, a_2, \ldots, a_n$ be given real numbers, which are not all zero. Determine the least value of

$$x_1^2 + \ldots + x_n^2 \; ,$$

where $x_1, \ldots, x_n$ are real numbers satisfying

$$a_1 x_1 + \ldots + a_n x_n = 1 \; .$$

Solution: We have, using Cauchy's inequality,

$$1 = |a_1 x_1 + \ldots + a_n x_n| \leq \left(\sum_{i=1}^{n} a_i^2 \right)^{1/2} \left(\sum_{i=1}^{n} x_i^2 \right)^{1/2} \; ,$$

so that

$$\sum_{i=1}^{n} x_i^2 \geq \frac{1}{\sum\limits_{i=1}^{n} a_i^2} \quad .$$

If we choose

$$x_i = \frac{a_i}{\sum\limits_{i=1}^{n} a_i^2} \quad (i = 1, 2, \ldots, n) \; ,$$

we have

$$\sum_{i=1}^{n} a_i x_i = 1$$

and

$$\sum_{i=1}^{n} x_i^2 = \frac{1}{\sum_{i=1}^{n} a_i^2} ,$$

so the minimum value of $\sum_{i=1}^{n} x_i^2$ subject to $\sum_{i=1}^{n} a_i x_i = 1$ is $\dfrac{1}{\sum_{i=1}^{n} a_i^2}$.

16. Evaluate the infinite series

$$S = 1 - \frac{2^3}{1!} + \frac{3^3}{2!} - \frac{4^3}{3!} + \cdots .$$

Solution: We have

$$(n+1)^3 \equiv n(n-1)(n-2) + 6n(n-1) + 7n + 1$$

so that

$$S = \sum_{n=0}^{\infty} \frac{(-1)^n (n+1)^3}{n!}$$

$$= \sum_{n=0}^{\infty} (-1)^n \frac{(n(n-1)(n-2) + 6n(n-1) + 7n + 1)}{n!}$$

$$= \sum_{n=3}^{\infty} \frac{(-1)^n}{(n-3)!} + 6 \sum_{n=2}^{\infty} \frac{(-1)^n}{(n-2)!} + 7 \sum_{n=1}^{\infty} \frac{(-1)^n}{n!} + \sum_{n=0}^{\infty} \frac{(-1)^n}{n!}$$

$$= -\sum_{m=0}^{\infty} \frac{(-1)^m}{m!} + 6 \sum_{m=0}^{\infty} \frac{(-1)^m}{m!} - 7 \sum_{m=0}^{\infty} \frac{(-1)^m}{m!} + \sum_{m=0}^{\infty} \frac{(-1)^m}{m!}$$

$$= -\sum_{m=0}^{\infty} \frac{(-1)^m}{m!} = -e^{-1} .$$

17. $F(x)$ is a differentiable function such that $F'(a-x) = F'(x)$ for all x satisfying $0 \leq x \leq a$. Evaluate $\int_0^a F(x)dx$ and give an example of such a function $F(x)$.

Solution: As
$$F'(a-x) = F'(x) , \quad 0 \leq x \leq a ,$$
we have by integrating
$$-F(a-x) = F(x) + C ,$$
where C is a constant. Taking $x = 0$ we obtain $C = -F(0) - F(a)$, so that
$$F(x) + F(a-x) = F(0) + F(a) .$$
Integrating again we get
$$\int_0^a F(x)\, dx + \int_0^a F(a-x)\, dx = a(F(0) + F(a)) .$$
As
$$\int_0^a F(a-x)\, dx = \int_0^a F(x)\, dx ,$$
the desired integral has the value $\frac{a}{2}(F(0) + F(a))$.

Two examples of such functions are
$$k \cos \frac{\pi x}{a} \quad \text{and} \quad k(2x^3 - 3ax^2) ,$$
where k is an arbitrary constant.

18. (a) Let r,s,t,u be the roots of the quartic equation
$$x^4 + Ax^3 + Bx^2 + Cx + D = 0 .$$
Prove that if $rs = tu$ then $A^2 D = C^2$.

(b) Let   a,b,c,d   be the roots of the quartic equation

$$y^4 + py^2 + qy + r = 0 .$$

Use (a) to determine the cubic equation (in terms of p,q,r) whose roots are

$$\frac{ab - cd}{a + b - c - d} , \quad \frac{ac - bd}{a + c - b - d} , \quad \frac{ad - bc}{a + d - b - c} .$$

<u>Solution:</u>   (a) As   r,s,t,u   are the roots of the quartic equation
$x^4 + Ax^3 + Bx^2 + Cx + D = 0$, we have

$$r + s + t + u = -A ,$$
$$rst + rsu + rtu + stu = -C ,$$
$$rstu = D .$$

Since   rs = tu   we have

$$\begin{aligned}
A^2 D &= (r + s + t + u)^2 r^2 s^2 \\
&= (r^2 s + rs^2 + rst + rsu)^2 \\
&= (rtu + stu + rst + rsu)^2 \\
&= C^2 .
\end{aligned}$$

(b) As the roots of the equation

$$y^4 + py^2 + qy + r = 0$$

are   a,b,c,d , we have

(18.1) $a + b + c + d = 0$,

(18.2) $ab + ac + ad + bc + bd + cd = p$,

(18.3) $abc + abd + acd + bcd = -q$,

(18.4) $abcd = r$.

Let   z   be a real or complex number. We begin by finding the quartic
equation whose roots are   a-z, b-z, c-z, d-z .

From (18.1) , we obtain

(18.5) $(a-z) + (b-z) + (c-z) + (d-z) = -4z$.

Similarly, from (18.1) and (18.2), we obtain

$(a-z)(b-z) + (a-z)(c-z) + (a-z)(d-z) + (b-z)(c-z) + (b-z)(d-z) + (c-z)(d$

$\qquad = (ab + ac + ad + bc + bd + cd) - 3(a + b + c + d)z + 6z^2$,

that is

(18.6) $(a-z)(b-z) + (a-z)(c-z) + \ldots + (c-z)(d-z) = p + 6z^2$.

  Next, from (18.1), (18.2) and (18.3), we have

  $(a-z)(b-z)(c-z)+(a-z)(b-z)(d-z)+(a-z)(c-z)(d-z)+(b-z)(c-z)(d-z)$

$= (abc+abd+acd+bcd) - 2(ab+ac+ad+bc+bd+cd)z + 3(a+b+c+d)z^2 - 4z^3$,

so that

$\qquad\qquad (a-z)(b-z)(c-z) + \ldots + (b-z)(c-z)(d-z)$

(18.7)

$\qquad\qquad\qquad = -q - 2pz - 4z^3$.

Also, from (18.1), (18.2), (18.3), (18.4), we have

  $(a-z)(b-z)(c-z)(d-z)$

$\qquad = abcd - (abc+\ldots+bcd)z + (ab+\ldots+cd)z^2 - (a+b+c+d)z^3 + z^4$,

so that

(18.8) $(a-z)(b-z)(c-z)(d-z) = r + qz + pz^2 + z^4$.

  Hence the desired quartic equation, whose roots are

$$a-z , \quad b-z , \quad c-z , \quad d-z ,$$

is

$\qquad y^4 + 4zy^3 + (p+6z^2)y^2 + (q+2pz+4z^3)y + (r+qz+pz^2+z^4) = 0$.

  To finish the problem we take $z_1 = \dfrac{ab - cd}{a + b - c - d}$, so that

$$(a - z_1)(b - z_1) = (c - z_1)(d - z_1) ,$$

and thus by (a) we have

$$16z_1^2(r + qz_1 + pz_1^2 + z_1^4) = (q + 2pz_1 + 4z_1^3)^2 ,$$

so that z_1 is a root of

(18.9) $$8qz^3 + 4(4r - p^2)z^2 - 4pqz - q^2 = 0 .$$

Similarly $z_2 = \dfrac{ac - bd}{a + c - b - d}$ and $z_3 = \dfrac{ad - bc}{a + d - b - c}$ are also roots of (18.9), which is the required cubic equation.

19. Let $p(x)$ be a monic polynomial of degree $m \geq 1$, and set

$$f_n(x) = e^{p(x)}D^n(e^{-p(x)}) ,$$

where n is a non-negative integer and $D \equiv \dfrac{d}{dx}$ denotes differentiation with respect to x.

Prove that $f_n(x)$ is a polynomial in x of degree $(mn - n)$. Determine the ratio of the coefficient of x^{mn-n} in $f_n(x)$ to the constant term in $f_n(x)$.

Solution: Differentiating $f_n(x)$ by the product rule, we obtain

$$f_n'(x) = e^{p(x)}D^{n+1}(e^{-p(x)}) + p'(x)e^{p(x)}D^n(e^{-p(x)}) ,$$

so that

$$f_n'(x) = f_{n+1}(x) + p'(x)f_n(x) ,$$

and so

(19.1) $$f_{n+1}(x) = f_n'(x) - p'(x)f_n(x) .$$

Clearly $f_0(x) = 1$ is a polynomial of degree 0, $f_1(x) = -p'(x)$ is a polynomial of degree $m-1$, and $f_2(x) = -p''(x) + p'(x)^2$ is a polynomial of degree $2m-2$. With the inductive hypothesis that $f_n(x)$ is a polynomial of degree $mn-n$, we easily deduce from (19.1) that

$f_{n+1}(x)$ is a polynomial of degree $m(n+1) - (n+1)$, and hence the principle of mathematical induction implies that $f_n(x)$ is a polynomial of degree $(mn-n)$ for all n .

Setting

$$p(x) = x^m + p_{m-1}x^{m-1} + \ldots + p_0$$

and

$$f_n(x) = a_{mn-n}x^{mn-n} + a_{mn-n-1}x^{mn-n-1} + \ldots + a_0 \, ,$$

we obtain, from (19.1),

$$a_{mn+m-n-1}x^{mn+m-n-1} + \ldots + a_0$$

$$= (mn-n)a_{mn-n}x^{mn-n-1} + (mn-n-1)a_{mn-n-1}x^{mn-n-2} + \ldots + a_1$$

$$- (mx^{m-1} + (m-1)p_{m-1}x^{m-2} + \ldots + p_1)$$

$$(a_{mn-n}x^{mn-n} + a_{mn-n-1}x^{mn-n-1} + \ldots + a_0) \, .$$

Equating coefficients of $x^{mn+m-n-1}$, we obtain

$$a_{mn+m-n-1} = -ma_{mn-n} \, .$$

Solving this recurrence relation, we obtain

$$a_{mn-n} = (-m)^n a_0 \, ,$$

that is

$$\frac{a_{mn-n}}{a_0} = (-m)^n \, .$$

20. Determine the real function of x whose power series is

$$\frac{x^3}{3!} + \frac{x^9}{9!} + \frac{x^{15}}{15!} + \ldots \, .$$

Solution: We make use of the complex cube of unity

$$w = \frac{-1 + \sqrt{-3}}{2} \quad ,$$

so that

(20.1) $w^3 = 1 \ , \quad w^2 + w + 1 = 0 \ .$

Now, for all real x , we have

$$\sinh x = x + \frac{x^3}{3!} + \frac{x^5}{5!} + \frac{x^7}{7!} + \frac{x^9}{9!} + \dots \ ,$$

$$\sinh wx = wx + \frac{x^3}{3!} + \frac{w^2 x^5}{5!} + \frac{w x^7}{7!} + \frac{x^9}{9!} + \dots \ ,$$

$$\sinh w^2 x = w^2 x + \frac{x^3}{3!} + \frac{w x^5}{5!} + \frac{w^2 x^7}{7!} + \frac{x^9}{9!} + \dots \ .$$

Adding these equations and using (20.1), we obtain

$$\sinh x + \sinh wx + \sinh w^2 x = 3\left(\frac{x^3}{3!} + \frac{x^9}{9!} + \dots \right) \ .$$

Now

$$\sinh wx = \sinh\left(\frac{-x}{2} + \frac{ix\sqrt{3}}{2} \right) = \sinh\left(\frac{-x}{2} \right)\cosh\left(\frac{ix\sqrt{3}}{2} \right) + \cosh\left(\frac{-x}{2} \right)\sinh\left(\frac{ix\sqrt{3}}{2} \right)$$

$$= -\sinh\left(\frac{x}{2} \right)\cos\left(\frac{x\sqrt{3}}{2} \right) + i\cosh\left(\frac{x}{2} \right)\sin\left(\frac{x\sqrt{3}}{2} \right) \ ,$$

and similarly

$$\sinh w^2 x = -\sinh\left(\frac{x}{2} \right)\cos\left(\frac{x\sqrt{3}}{2} \right) - i\cosh\left(\frac{x}{2} \right)\sin\left(\frac{x\sqrt{3}}{2} \right) \ ,$$

and so

$$\sinh wx + \sinh w^2 x = -2\sinh\left(\frac{x}{2} \right)\cos\left(\frac{x\sqrt{3}}{2} \right) \ ,$$

giving

$$\frac{x^3}{3!} + \frac{x^9}{9!} + \dots = \frac{2}{3}\sinh\left(\frac{x}{2} \right)\left(\cosh\left(\frac{x}{2} \right) - \cos\left(\frac{x\sqrt{3}}{2} \right) \right) \quad .$$

21. Determine the value of the integral

(21.0) $$I_n = \int_0^\pi \left(\frac{\sin nx}{\sin x}\right)^2 dx \; ,$$

for all positive integral values of n .

<u>Solution</u>: We will show that $I_n = n\pi$, $n = 1,2,3,\ldots$.
From (21.0), we have for $n \geq 2$

$$D_n = I_n - I_{n-1} = \int_0^\pi \frac{(\sin^2 nx - \sin^2 (n-1)x)}{\sin^2 x} \, dx$$

$$= \int_0^\pi \frac{(\sin nx - \sin (n-1)x)(\sin nx + \sin (n-1)x)}{\sin^2 x} dx$$

$$= \int_0^\pi \frac{2 \sin \frac{x}{2} \cos(nx - \frac{x}{2}) \; 2 \sin(nx - \frac{x}{2}) \cos \frac{x}{2}}{\sin^2 x} dx$$

$$= \int_0^\pi \frac{\sin x \cdot \sin (2n-1)x}{\sin^2 x} \, dx$$

$$= \int_0^\pi \frac{\sin (2n-1)x}{\sin x} \, dx \; ,$$

that is

(21.1) $$D_n = J_{2n-1} \; , \quad n \geq 2 \; ,$$

where

$$J_m = \int_0^\pi \frac{\sin mx}{\sin x} \, dx \; , \quad m = 0,1,2,\ldots \; .$$

Now, for $m \geq 2$, we have

$$J_m - J_{m-2} = \int_0^\pi \frac{(\sin mx - \sin (m-2)x)}{\sin x} \, dx$$

$$= \int_0^\pi \frac{2 \sin x \cos (m-1)x}{\sin x} \, dx$$

$$= 2 \int_0^\pi \cos(m-1)x \, dx$$

$$= 2 \left[\frac{\sin(m-1)x}{m-1} \right]_0^\pi$$

$$= 0 \ ,$$

so that

$$J_m = J_{m-2} = J_{m-4} = \cdots = \begin{cases} J_0 = 0 \ , & \text{if } m \text{ even,} \\ J_1 = \pi \ , & \text{if } m \text{ odd.} \end{cases}$$

Hence, from (21.1), we obtain $D_n = \pi$, $n \geq 2$, so that $I_n = n\pi$, $n \geq 1$, as $I_1 = \pi$.

22. During the year 1985, a convenience store, which was open 7 days a week, sold at least one book each day, and a total of 600 books over the entire year. Must there have been a period of consecutive days when exactly 129 books were sold?

Solution: Let a_i , $i = 1,2,3,\ldots,365$, denote the number of books sold by the store during the period from the first day to the ith day inclusive, so that

$$1 \leq a_1 < a_2 < \cdots < a_{365} = 600 \ ,$$

and thus

$$130 \leq a_1 + 129 < a_2 + 129 < \cdots < a_{365} + 129 = 729 \ .$$

Hence $a_1, \ldots, a_{365}, a_1 + 129, \ldots, a_{365} + 129$ are 730 positive integers between 1 and 729 inclusive. Thus, by Dirichlet's box principle, two of these numbers must be the same. As $a_1, \ldots, a_{365}$ are all distinct and $a_1 + 129, \ldots, a_{365} + 129$ are all distinct, one of the a_i

must be the same as one of the $a_i + 129$, say,

$$a_k = a_\ell + 129 , \quad 1 \leq \ell < k \leq 365 .$$

Hence $a_k - a_\ell = 129$ and so 129 books were sold between the $(\ell+1)^{th}$ day and the kth day inclusive.

23. Find a polynomial $f(x,y)$ with rational coefficients such that as m and n run through all positive integral values, $f(m,n)$ takes on all positive integral values once and once only.

<u>Solution</u>: For any positive integer k we can define a unique pair of integers (r,m) by

$$\begin{cases} \dfrac{(r-1)(r-2)}{2} < k \leq \dfrac{r(r-1)}{2} , \\[2mm] m = k - \dfrac{(r-1)(r-2)}{2} . \end{cases}$$

Clearly we have

$$0 < m \leq \frac{r(r-1)}{2} - \frac{(r-1)(r-2)}{2} = r-1 ,$$

that is

$$1 \leq m < r ,$$

so that r and m are positive integers. Moreover, we can define a positive integer n uniquely by $r = m + n$, which gives

$$k = \frac{(m+n-1)(m+n-2)}{2} + m ,$$

and a polynomial of the required type is therefore

$$f(x,y) = \frac{(x+y-1)(x+y-2)}{2} + x .$$

24. Let m be a positive squarefree integer. Let R,S be positive integers. Give a condition involving R,S,m which guarantees that there do not exist rational numbers x,y,z and w such that

(24.0) $R + 2S\sqrt{m} = (x + y\sqrt{m})^2 + (z + w\sqrt{m})^2$.

Solution: If there exist rational numbers x,y,z and w such that (24.0) holds then

$$R - 2S\sqrt{m} = (x-y\sqrt{m})^2 + (z-w\sqrt{m})^2 \geq 0 ,$$

and so a condition that will guarantee the non-solvability of (24.0) is $R - 2S\sqrt{m} < 0$, that is

$$\frac{R}{2S} < \sqrt{m} .$$

25. Let k and h be integers with $1 \leq k < h$. Evaluate the limit

$$L = \lim_{n \to \infty} \prod_{r=kn+1}^{hn} \left(1 - \frac{r}{n^2}\right) .$$

Solution: For $|x| < 1$ we have

$$\ln (1 - x) = - \sum_{s=1}^{\infty} \frac{x^s}{s} ,$$

and so

$$x + \ln (1 - x) = - \sum_{s=2}^{\infty} \frac{x^s}{s} ,$$

giving

$$\left| x + \ln (1 - x) \right| = \left| \sum_{s=2}^{\infty} \frac{x^s}{s} \right|$$

$$\leq \sum_{s=2}^{\infty} \frac{|x|^s}{s}$$

$$\leq \frac{1}{2} \sum_{s=2}^{\infty} |x|^s$$

$$= \frac{|x|^2}{2(1-|x|)} \quad .$$

Taking $x = \frac{r}{n^2}$ $\quad (kn < r \leq hn)$ we obtain

$$\left| \frac{r}{n^2} + \ell n \left(1 - \frac{r}{n^2}\right) \right| \leq \frac{r^2}{2n^4\left(1 - \frac{r}{n^2}\right)}$$

$$\leq \frac{r^2}{2n^4\left(1 - \frac{h}{n}\right)} \quad .$$

Thus we obtain

$$\left| \sum_{r=kn+1}^{hn} \left(\frac{r}{n^2} + \ell n \left(1 - \frac{r}{n^2}\right) \right) \right| \leq \frac{1}{2n^4\left(1-\frac{h}{n}\right)} \sum_{r=kn+1}^{hn} r^2$$

$$\leq \frac{1}{2n^4\left(1-\frac{h}{n}\right)} \sum_{r=1}^{hn} r^2$$

$$< \frac{h^3 n^3}{2n^4\left(1-\frac{h}{n}\right)}$$

$$= \frac{h^3}{2n\left(1 - \frac{h}{n}\right)}$$

$$\rightarrow 0 \quad \text{as} \quad n \rightarrow \infty \quad ,$$

showing that

(25.1) $\qquad \lim_{n \to \infty} \sum_{r=kn+1}^{hn} \left[\frac{r}{n^2} + \ell n \left(1 - \frac{r}{n^2}\right) \right] = 0 \quad .$

Next we have

$$\sum_{r=kn+1}^{hn} \frac{r}{n^2} = \frac{1}{n^2} \left\{ \frac{hn(hn+1)}{2} - \frac{kn(kn+1)}{2} \right\} = \frac{(h^2-k^2)}{2} + \frac{(h-k)}{2n} \quad ,$$

so that

(25.2) $$\lim_{n \to \infty} \sum_{r=kn+1}^{hn} \frac{r}{n^2} = \frac{(h^2-k^2)}{2} \; .$$

Thus from (25.1) and (25.2) we obtain

$$\lim_{n \to \infty} \ell n \prod_{r=kn+1}^{hn} (1 - \frac{r}{n^2}) = -\frac{1}{2}(h^2-k^2)$$

and so

$$\lim_{n \to \infty} \prod_{r=kn+1}^{hn} (1 - \frac{r}{n^2}) = e^{-\frac{1}{2}(h^2-k^2)} \; .$$

26. Let $f(x)$ be a continuous function on $[0,a]$ such that $f(x)f(a-x) = 1$, where $a > 0$. Prove that there exist infinitely many such functions $f(x)$, and evaluate

$$\int_0^a \frac{dx}{1 + f(x)} \; .$$

Solution: The function $f(x) = e^{x - \frac{a}{2}}$ is continuous for all x and satisfies $f(x)f(a-x) = 1$, so that $f(x)^n$ $(n = 0,1,2,\ldots)$ gives an infinite family of functions of the required type.

Setting $x = a - y$ we obtain

$$I = \int_0^a \frac{dx}{1 + f(x)} = \int_a^0 \frac{-dy}{1 + f(a-y)}$$

$$= \int_0^a \frac{dy}{1 + f(a-y)} = \int_0^a \frac{f(y)dy}{f(y) + 1}$$

$$= \int_0^a \left\{ 1 - \frac{1}{1 + f(y)} \right\} dy$$

$$= a - \int_0^a \frac{dy}{1 + f(y)}$$

$$= a - I \ , \quad \text{so that} \quad I = \frac{a}{2} \ .$$

27. The positive numbers $a_1, a_2, a_3, \ldots$ satisfy

(27.0)
$$\sum_{r=1}^{n} a_r^3 = \left(\sum_{r=1}^{n} a_r \right)^2 , \quad n = 1, 2, 3, \ldots \ .$$

Is it true that $a_r = r$ for $r = 1, 2, 3, \ldots$?

Solution: The answer is yes. We proceed by mathematical induction, making use of the identity

$$1^3 + 2^3 + \ldots + n^3 = (1 + 2 + \ldots + n)^2 \ .$$

Taking $n = 1$ in (27.0) gives $a_1^3 = a_1^2$, which means that $a_1 = 1$ because $a_1 > 0$.

Next, assume that $a_k = k$ for $k = 1, 2, \ldots, n-1$. The equation (27.0) gives

$$1^3 + 2^3 + \ldots + (n-1)^3 + a_n^3 = (1 + 2 + \ldots + (n-1) + a_n)^2$$

so that

$$(1+2+\ldots+(n-1))^2 + a_n^3 = (1+2+\ldots+(n-1))^2 + 2(1+2+\ldots+(n-1))a_n + a_n^2 ,$$

that is,

(27.1)
$$a_n^3 = (n-1) n a_n + a_n^2 \ .$$

As $a_n > 0$, we see that (27.1) gives $a_n = n$, thus completing the inductive step.

28. Let $p > 0$ be a real number and let n be a non-negative integer. Evaluate

$$(28.0) \qquad u_n(p) = \int_0^\infty e^{-px} \sin^n x \, dx \ .$$

<u>Solution:</u> For $n \geq 2$ and $p > 0$, integrating $u_n = u_n(p)$ by parts we obtain

$$u_n = -\frac{1}{p} e^{-px} \sin^n x \Big|_0^\infty + \int_0^\infty n \sin^{n-1} x \cos x \cdot \frac{e^{-px}}{p} \, dx$$

$$= \frac{n}{p} \int_0^\infty \sin^{n-1} x \cos x \, e^{-px} \, dx \ .$$

Integrating by parts again, we get

$$u_n = \frac{n}{p} \left\{ -\frac{1}{p} e^{-px} \sin^{n-1} x \cos x \Big|_0^\infty \right.$$

$$\left. + \int_0^\infty \left((n-1) \sin^{n-2} x \cos^2 x - \sin^n x \right) \frac{e^{-px}}{p} \, dx \right\}$$

$$= \frac{n}{p^2} \int_0^\infty \left[(n-1) \sin^{n-2} x \, (1 - \sin^2 x) - \sin^n x \right] e^{-px} \, dx$$

$$= \frac{n(n-1)}{p} \int_0^\infty e^{-px} \sin^{n-2} x \, dx - \frac{n^2}{p^2} \int_0^\infty e^{-px} \sin^n x \, dx \ ,$$

that is

$$u_n = \frac{n(n-1)}{p} u_{n-2} - \frac{n^2}{p^2} u_n \ , \qquad n \geq 2 \ .$$

Thus we have

$$u_n = \frac{n(n-1)}{n^2 + p^2} u_{n-2} \ , \qquad n \geq 2 \ ,$$

giving

$$
u_n = \begin{cases}
\dfrac{n(n-1)}{n^2+p^2} \cdot \dfrac{(n-2)(n-3)}{(n-2)^2+p^2} \cdots \dfrac{2 \cdot 1}{2^2+p^2} u_0 & , \text{ if } n \text{ even,} \\[4mm]
\dfrac{n(n-1)}{\ddot{n}^2+p^2} \cdot \dfrac{(n-2)(n-3)}{(n-2)^2+p^2} \cdots \dfrac{3 \cdot 2}{3^2+p^2} u_1 & , \text{ if } n \text{ odd .}
\end{cases}
$$

One easily sees that $u_0 = \dfrac{1}{p}$ and $u_1 = \dfrac{1}{1+p^2}$, so that

$$
u_n = \begin{cases}
\dfrac{n!}{p \displaystyle\prod_{i=1}^{n/2} ((2i)^2 + p^2)} & , \text{ if } n \text{ even,} \\[8mm]
\dfrac{n!}{\displaystyle\prod_{i=0}^{(n-1)/2} ((2i+1)^2 + p^2)} & , \text{ if } n \text{ odd.}
\end{cases}
$$

29. Evaluate

(29.0)
$$
\sum_{r=0}^{n-2} 2^r \tan \frac{\pi}{2^{n-r}} ,
$$

for integers $n \geq 2$.

<u>Solution</u>: We use the identity

$$
\tan A = \cot A - 2 \cot 2A ,
$$

which is easily verified as

$$
2 \cot 2A = \frac{2 \cos 2A}{\sin 2A} = \frac{\cos^2 A - \sin^2 A}{\sin A \cos A} = \cot A - \tan A .
$$

Then we have

$$
\sum_{r=0}^{n-2} 2^r \tan \frac{\pi}{2^{n-r}} = \sum_{r=0}^{n-2} 2^r \left(\cot \frac{\pi}{2^{n-r}} - 2 \cot \frac{\pi}{2^{n-r-1}} \right)
$$

$$= \sum_{r=0}^{n-2} 2^r \cot \frac{\pi}{2^{n-r}} - \sum_{r=0}^{n-2} 2^{r+1} \cot \frac{\pi}{2^{n-r-1}}$$

$$= \sum_{r=0}^{n-2} 2^r \cot \frac{\pi}{2^{n-r}} - \sum_{r=1}^{n-1} 2^r \cot \frac{\pi}{2^{n-r}}$$

$$= 2^0 \cot \frac{\pi}{2^n} - 2^{n-1} \cot \frac{\pi}{2^{n-(n-1)}}$$

$$= \cot \frac{\pi}{2^n} - 2^{n-1} \cot \frac{\pi}{2}$$

$$= \cot \frac{\pi}{2^n} \quad .$$

Second solution (due to L. Smith):

For $0 < B < 2\pi$ we consider the integral

$$I(B) = \int_0^B \left(\sum_{r=0}^{n-2} 2^r \tan \frac{A}{2^{n-r}} \right) dA$$

$$= \sum_{r=0}^{n-2} 2^r \int_0^B \tan \frac{A}{2^{n-r}} dA$$

$$= \sum_{r=0}^{n-2} 2^r \cdot 2^{n-r} \, \ell n \cos \frac{A}{2^{n-r}} \bigg|_0^B$$

$$= 2^n \sum_{r=0}^{n-2} \ell n \cos \frac{B}{2^{n-r}}$$

$$= 2^n \ell n \prod_{r=0}^{n-2} \cos \frac{B}{2^{n-r}}$$

$$= 2^n \ell n \left(\frac{\sin \frac{B}{2}}{2^{n-1} \sin \frac{B}{2^n}} \right)$$

$$= 2^n \, \ell n \, \sin \frac{B}{2} - 2^n \, \ell n \, \sin \frac{B}{2^n} - 2^n \, \ell n \, 2^{n-1} \; .$$

Differentiating we obtain

$$I'(B) = \cot \frac{B}{2^n} - 2^{n-1} \cot \frac{B}{2} \; .$$

Taking $B = \pi$ we get

$$\sum_{r=0}^{n-2} 2^r \tan \frac{\pi}{2^{n-r}} = \cot \frac{\pi}{2^n} \; .$$

30. Let $n \geq 2$ be an integer. A selection $\{s = a_i : i=1,2,\ldots,k\}$ of k $(2 \leq k \leq n)$ elements from the set $N = \{1,2,3,\ldots,n\}$ such that $a_1 < a_2 < \ldots < a_k$ is called a k-selection. For any k-selection S, define

$$W(S) = \min \{a_{i+1} - a_i : i = 1,2,\ldots,k-1\} \; .$$

If a k-selection S is chosen at random from N, what is the probability that

$$W(S) = r \; ,$$

where r is a natural number?

Solution: Let $f_r(n,k)$, $2 \leq k \leq n$, denote the number of k-selections S from N such that $W(S) \geq r$, $r = 1,2,3,\ldots$. We will show that

$$(30.1) \qquad f_r(n,k) = \binom{n-(k-1)(r-1)}{k} \; , \qquad r = 1,2,3,\ldots,$$

where $\binom{m}{k} = 0$ for any integer $m < k$. When $r = 1$, $f_1(n,k)$ enumerates all k-selections S from N, so that $f_1(n,k) = \binom{n}{k}$, and hence (30.1) holds in this case. Now suppose $r \geq 2$. Let $S = \{a_1, a_2, \ldots, a_k\}$ be a k-selection from N with $W(S) \geq r$,

so that $a_i + r \leq a_{i+1}$ for $i = 1,2,\ldots,k-1$. The mapping F defined by

(30.2) $F(S) = \{a_1, a_2 - (r-1), a_3 - 2(r-1), \ldots, a_k - (k-1)(r-1)\}$

associates with a k-selection S from N having $W(S) \geq r$, a k-selection from the set

$$M = \{1, 2, \ldots, n - (k-1)(r-1)\} .$$

Clearly F is one-to-one and onto, so that $f_r(n,k)$ is just the number of k-selections from M, which is given by the right side of (30.1).

Thus the required probability is

$$\frac{f_r(n,k) - f_{r+1}(n,k)}{f_1(n,k)} = \frac{\binom{n-(k-1)(r-1)}{k} - \binom{n-(k-1)r}{k}}{\binom{n}{k}} .$$

31. Let $k \geq 2$ be a fixed integer. For $n = 1,2,3,\ldots$ define

$$a_n = \begin{cases} 1 & , \text{ if } n \text{ is not a multiple of } k , \\ -(k-1) & , \text{ if } n \text{ is a multiple of } k . \end{cases}$$

Evaluate the series $\displaystyle\sum_{n=1}^{\infty} \frac{a_n}{n}$.

Solution: Let s_n be the sum of the first n terms of the given series. For each $n \geq 1$ we have uniquely

$$n = kq_n + r_n , \quad 0 \leq r_n < k ,$$

and since

$$-(k-1)/tk = 1/tk - 1/t$$

for $t = 1,2,\ldots,q_n$, we have

$$s_n = \left(1 + \frac{1}{2} + \ldots + \frac{1}{n}\right) - \left(1 + \frac{1}{2} + \ldots + \frac{1}{q_n}\right) \quad .$$

Now, $n = q_n(k + r_n/q_n)$ so

$$\ell n \ n = \ell n \ q_n + \ell n \ (k + r_n/q_n) \ ,$$

and hence

$$s_n = (1 + \frac{1}{2} + \ldots + \frac{1}{n} - \ell n \ n) - (1 + \frac{1}{2} + \ldots + \frac{1}{q_n} - \ell n \ q_n) + \ell n \ (k + r_n/q_n)$$

$$= u_n - v_n + w_n \ .$$

The sequence $\{u_n: n = 1,2,3,\ldots\}$ converges to Euler's constant c; the sequence $\{v_n: n = 1,2,3,\ldots\}$ also converges to c as $n \to \infty$ implies $q_n \to \infty$; and the sequence $\{w_n: n = 1,2,3,\ldots\}$ converges to $\ell n \ k$ as r_n is bounded. Thus, we have $\sum\limits_{n=1}^{\infty} \frac{a_n}{n} = \ell n \ k$.

32. Prove that

$$\int_0^\infty x^m e^{-x} \sin x \ dx = \frac{m!}{2^{(m+2)/2}} \sin (m+1)\pi/4$$

for $m = 0,1,2,\ldots$.

Solution: We set for $m = 0,1,2,\ldots$

$$S_m = \int_0^\infty x^m e^{-x} \sin x \ dx \ ,$$

$$C_m = \int_0^\infty x^m e^{-x} \cos x \ dx \ ,$$

$$E_m = \int_0^\infty x^m e^{(i-1)x} \ dx \ .$$

As

$$e^{ix} = \cos x + i \sin x ,$$

we have

$$E_m = C_m + i S_m .$$

Integrating E_m $(m \geq 1)$ by parts, we obtain

$$E_m = x^m \frac{e^{(i-1)x}}{i-1} \Big|_0^\infty - \int_0^\infty m \, x^{m-1} \frac{e^{(i-1)x}}{i-1} \, dx$$

$$= \frac{-m}{i-1} E_{m-1} , \quad m \geq 1 .$$

Hence we have

$$E_m = \frac{(-1)^m m!}{(i-1)^m} E_0 , \quad m \geq 0 .$$

Clearly $E_0 = -\frac{1}{i-1}$, so that for $m \geq 0$

$$E_m = \frac{(-1)^{m+1} m!}{(i-1)^{m+1}} = \frac{m! \, (i+1)^{m+1}}{2^{m+1}} .$$

Finally, we obtain for $m \geq 0$

$$S_m = \frac{1}{2i}(E_m - \overline{E}_m)$$

$$= \frac{m!}{2^{m+2} i} \left\{ (1+i)^{m+1} - (1-i)^{m+1} \right\} ,$$

and so, by Demoivre's theorem, we have

$$S_m = \frac{m!}{2^{m+2} i} \left\{ 2^{\frac{m+1}{2}} (\cos(m+1)\frac{\pi}{4} + i \sin(m+1)\frac{\pi}{4}) - 2^{\frac{m+1}{2}} (\cos(m+1)\frac{\pi}{4} - i \sin(m+1)\frac{\pi}{4}) \right\}$$

$$= \frac{m!}{2^{(m+1)/2}} \sin(m+1)\frac{\pi}{4} , \quad m \geq 0 .$$

33. For a real number u set

(33.0) $I(u) = \int_0^\pi \ell n(1 - 2u \cos x + u^2) \, dx$.

Prove that

$$I(u) = I(-u) = \frac{1}{2} I(u^2) ,$$

and hence evaluate I(u) for all values of u .

Solution: We will show that

(33.1) $I(u) = \begin{cases} 0 & , \text{ if } |u| \leq 1 , \\ 2\pi \ell n |u| & , \text{ if } |u| > 1 . \end{cases}$

First, we prove that

(33.2) $I(u) = I(-u)$.

Setting $x = \pi - y$ in (33.0), we obtain

$$I(u) = \int_0^\pi \ell n(1 + 2u \cos y + u^2) \, dy$$

$$= I(-u) .$$

Next we show that

(33.3) $I(u) + I(-u) = I(u^2)$.

We have

$$(1 - 2u \cos x + u^2)(1 + 2u \cos x + u^2) = (1 + u^2)^2 - (2u \cos x)^2$$

$$= 1 + u^4 + 2u^2(1 - 2\cos^2 x)$$

$$= 1 - 2u^2 \cos 2x + u^4 ,$$

so that

$$\ell n(1 - 2u \cos x + u^2) + \ell n(1 + 2u \cos x + u^2) = \ell n(1 - 2u^2 \cos 2x + u^4) ,$$

and thus

$$I(u) + I(-u) = \int_0^\pi \ell n(1 - 2u^2 \cos 2x + u^4) \, dx ,$$

and setting y = 2x , we obtain

$$I(u) + I(-u) = \frac{1}{2} \int_0^{2\pi} \ell n(1 - 2u^2 \cos y + u^4) \, dy$$

$$= \frac{1}{2} I(u^2) + \frac{1}{2} \int_\pi^{2\pi} \ell n(1 - 2u^2 \cos y + u^4) \, dy \ .$$

Setting y = 2π - z in the last integral, we obtain

$$\int_\pi^{2\pi} \ell n(1 - 2u^2 \cos y + u^4) \, dy = I(u^2) \ ,$$

proving (33.3) as required.

From (33.2) and (33.3), we deduce

(33.4) $I(u) = I(-u) = \frac{1}{2} I(u^2) \ .$

For $|u| = 1$, that is u = ±1, we have

$$I(1) = I(-1) = \frac{1}{2} I(1) \ ,$$

so

$$I(1) = I(-1) = 0 \ .$$

For $|u| < 1$ we have

$$I(u) = \frac{1}{2} I(u^2) = \frac{1}{2^2} I(u^4) = \frac{1}{2^3} I(u^8) = \ldots = \frac{1}{2^n} I(u^{2^n}) \ ,$$

for all positive integers n . Letting n → +∞ , we have $u^{2^n} \to 0$
as $|u| < 1$, and I(u) being continuous gives

$$\lim_{n \to \infty} I(u^{2^n}) = I(0) \ .$$

[In fact it follows from (33.4) that I(0) = 0 .] Hence, as $\frac{1}{2^n} \to 0$,
we obtain

$$\lim_{n \to \infty} \frac{1}{2^n} I(u^{2^n}) = 0 \ ,$$

giving

$$I(u) = 0 , \quad \text{for} \quad |u| < 1 .$$

Next, for $|u| > 1$, we set $u = \frac{1}{v}$, so that $0 < |v| < 1$. Then, for $u > 0$, we have

$$I(u) = \int_0^\pi \ell n(1 - \frac{2}{v}\cos x + \frac{1}{v^2}) \, dx$$

$$= \int_0^\pi \{\ell n(1 - 2v\cos x + v^2) - 2\ell n\, v\} \, dx$$

$$= I(v) - 2\ell n\, v \int_0^\pi dx$$

$$= 0 - 2\pi \ell n\, v$$

$$= 2\pi \ell n\, u .$$

Finally, if $u < 0$, we have

$$I(u) = I(-u) = 2\pi \ell n\, (-u) ,$$

so that for all u with $|u| > 1$ we have

$$I(u) = 2\pi \ell n\, |u| .$$

34. For each natural number $k \geq 2$ the set of natural numbers is partitioned into a sequence of sets $\{A_n(k) : n = 1,2,3,\dots \}$ as follows: $A_1(k)$ consists of the first k natural numbers, $A_2(k)$ consists of the next $k+1$ natural numbers, $A_3(k)$ consists of the next $k+2$ natural numbers, etc. The sum of the natural numbers in $A_n(k)$ is denoted by $s_n(k)$. Determine the least value of $n = n(k)$ such that $s_n(k) > 3k^3 - 5k^2$, for $k = 2,3,\dots$.

Solution: The last element in $A_n(k)$ is the number

$$k + (k+1) + (k+2) + \ldots + (k+n-1) = nk + (1+2+\ldots+(n-1))$$
$$= nk + \frac{(n-1)n}{2} .$$

Since there are $k+n-1$ numbers in $A_n(k)$, the first element in $A_n(k)$ is

$$\left[nk + \frac{(n-1)n}{2}\right] - (k+n-1) + 1 = (n-1)k + \frac{1}{2}(n^2 - 3n + 4) .$$

Hence we have

(34.1) $$s_n(k) = \frac{(k+n-1)}{2}\left((2n-1)k + (n^2 - 2n + 2)\right) .$$

Taking $k = 2,3,4,5$ in (34.1) suggests that $n = k$ may be the required value of n. To prove this conjecture we calculate $s_{k-1}(k)$ and $s_k(k)$. We have

$$s_{k-1}(k) = (k-1)(3k^2 - 7k + 5) = 3k^3 - 10k^2 + 12k - 5$$

and

$$s_k(k) = \frac{(2k-1)(3k^2 - 3k + 2)}{2} = 3k^3 - \frac{9}{2}k^2 + \frac{7}{2}k - 1 .$$

One easily checks that for $k = 2,3,\ldots$

$$s_k(k) > 3k^3 - 5k^2 > s_{k-1}(k) ,$$

so that $n(k) = k$ is the required minimal value.

35. Let $\{p_n : n = 1,2,3,\ldots\}$ be a sequence of real numbers such that $p_n \geq 1$ for $n = 1,2,3,\ldots$. Does the series

(35.0) $$\sum_{n=1}^{\infty} \frac{[p_n]-1}{([p_1]+1)([p_2]+1)\ldots([p_n]+1)}$$

converge?

Solution: The answer is yes.

$$\text{Let} \quad a_n = \frac{p_n - 1}{p_1 \cdots p_n} \, , \quad n = 1, 2, \ldots \, .$$

Then we have for $m \geq 1$,

$$S_m = \sum_{n=1}^{m} a_n = \sum_{n=1}^{m} \left(\frac{1}{p_1 \cdots p_{n-1}} - \frac{1}{p_1 \cdots p_n} \right)$$

$$= 1 - \frac{1}{p_1 \cdots p_m}$$

$$\leq 1 - \frac{1}{p_1 \cdots p_{m+1}}$$

$$= S_{m+1} \, ,$$

so that $\{S_m : m = 1, 2, 3, \ldots \}$ is an increasing sequence which is bounded above by 1 . Hence $\lim\limits_{m \to \infty} S_m$ exists, showing that $\sum\limits_{n=1}^{\infty} a_n$ converges. Finally, as

$$\frac{[p_n] - 1}{([p_1] + 1)([p_2] + 1) \ldots ([p_n] + 1)} \leq a_n \, , \quad n = 1, 2, 3, \ldots \, ,$$

the series given by (35.0) converges by the comparison test.

36. Let $f(x)$, $g(x)$ be polynomials with real coefficients of degrees $n+1$, n respectively, where $n \geq 0$, and with positive leading coefficients A , B respectively. Evaluate

$$L = \lim_{x \to \infty} g(x) \int_0^x e^{f(t) - f(x)} \, dx \, ,$$

in terms of A , B and n .

Solution: As $A > 0$ and $B > 0$, we have

$$\lim_{x \to \infty} \int_0^x e^{f(t)} \, dt = +\infty$$

and

$$\lim_{x \to \infty} \frac{e^{f(x)}}{g(x)} = +\infty \quad .$$

Moreover

$$\lim_{x \to \infty} \frac{\dfrac{d}{dx}\left(\int_0^x e^{f(t)} dt \right)}{\dfrac{d}{dx}\left(\dfrac{e^{f(x)}}{g(x)} \right)}$$

$$= \lim_{x \to \infty} \frac{g(x)^2}{f'(x)g(x) - g'(x)}$$

$$= \lim_{x \to \infty} \frac{B^2 x^{2n} + \ldots}{AB(n+1)x^{2n} + \ldots}$$

$$= \frac{B}{(n+1)A} \quad ,$$

so, by L'Hôpital's rule, we have

$$L = \lim_{x \to \infty} \frac{\int_0^x e^{f(t)} \, dt}{e^{f(x)}/g(x)}$$

$$= \lim_{x \to \infty} \frac{\dfrac{d}{dx}\left(\int_0^x e^{f(t)} dt \right)}{\dfrac{d}{dx}\left[e^{f(x)}/g(x) \right]}$$

$$= \frac{B}{(n+1)A} \quad .$$

37. The lengths of two altitudes of a triangle are h and k, where $h \neq k$. Determine upper and lower bounds for the length of the third altitude in terms of h and k.

Solution: We show that

(37.1) $\dfrac{hk}{h+k} < \ell < \dfrac{hk}{|h-k|}$.

Let the points P, Q, R be chosen on the sides BC, CA, AB (possibly extended) respectively of the triangle ABC so that AP, BQ, CR are the altitudes of the triangle. Set $a = |BC|$, $b = |CA|$, $c = |AB|$, $h = |AP|$, $k = |BQ|$, $\ell = |CR|$. Clearly

$$ah = bk = c\ell ,$$

so that

$$\frac{a}{c} = \frac{\ell}{h} , \quad \frac{b}{c} = \frac{\ell}{k} .$$

Without loss of generality we may suppose that $h < k$. From the inequality

$$a < b + c ,$$

we obtain

$$\frac{a}{c} < \frac{b}{c} + 1 , \quad \frac{\ell}{h} < \frac{\ell}{k} + 1 ,$$

so that

$$\ell\left(\frac{1}{h} - \frac{1}{k}\right) < 1 ,$$

that is

$$\ell < \frac{hk}{k-h} .$$

Also from the inequality

$$c < a + b ,$$

we obtain

$$1 < \frac{a}{c} + \frac{b}{c} , \quad 1 < \frac{\ell}{h} + \frac{\ell}{k} ,$$

that is

$$\ell > \frac{hk}{h+k} \ .$$

This completes the proof of (37.1).

38. Prove that

$$P_{n,r} = P_{n,r}(x) = \frac{(1-x^{n+1})(1-x^{n+2})\ldots(1-x^{n-r})}{(1-x)(1-x^2)\ldots(1-x^r)}$$

is a polynomial in x of degree nr , where n and r are non-negative integers. (When $r = 0$ the empty product is understood to be 1 and we have $P_{n,0} = 1$ for all $n \geq 0$.)

Solution: For $n \geq 0$ and $r \geq 1$ we have

$$P_{n+1,r} - x^r P_{n,r} = \frac{(1-x^{n+2})\ldots(1-x^{n+r+1})}{(1-x)\ldots(1-x^r)} - x^r \frac{(1-x^{n+1})\ldots(1-x^{n+r})}{(1-x)\ldots(1-x^r)}$$

$$= \frac{(1-x^{n+2})\ldots(1-x^{n+r})\left((1-x^{n+r+1}) - x^r(1-x^{n+1})\right)}{(1-x)\ldots(1-x^r)}$$

$$= \frac{(1-x^{n+2})\ldots(1-x^{n+r})}{(1-x)\ldots(1-x^{r-1})} \ ,$$

so that

(38.1) $P_{n+1,r} - x^r P_{n,r} = P_{n+1,r-1}$.

We now make the inductive hypothesis that $P_{n,r}$ is a polynomial of degree nr for all pairs (n,r) of non-negative integers satisfying $n + r \leq k$, where k is a non-negative integer. This is clearly true when $k = 0$, as, in this case, we must have $n = r = 0$, and $P_{0,0} = 1$. Now let (n,r) be a pair of non-negative integers such that $n + r = k + 1$. For $n \geq 1$ and $r \geq 1$, by (38.1), we have

(38.2) $$P_{n,r} = x^r P_{n-1,r} + P_{n,r-1} \ .$$

As $(n-1) + r = n + (r-1) = k$, by the inductive hypothesis, $P_{n-1,r}$ is a polynomial of degree $(n-1)r$ and $P_{n,r-1}$ is a polynomial of degree $n(r-1)$. Hence, by (38.2), $P_{n,r}$ is a polynomial of degree

$$\max(r + (n-1)r \ , \ n(r-1)) = nr \ .$$

$P_{n,r}$ is clearly a polynomial of degree 0 in the remaining cases $n = 0$ and $r = 0$. The result now follows by the principle of mathematical induction.

$39.$ Let A, B, C, D, E be integers such that $B \neq 0$ and

$$F = AD^2 - BCD + B^2E \neq 0 \ .$$

Prove that the number N of pairs of integers (x,y) such that

(39.0) $$Ax^2 + Bxy + Cx + Dy + E = 0 \ ,$$

satisfies

$$N \leq 2d(|F|) \ ,$$

where, for integers $n \geq 1$, $d(n)$ denotes the number of positive divisors of n .

Solution: Let (x,y) be a solution in integers of (39.0), so that

(39.1) $$-(Bx + D)y = (Ax^2 + Cx + E) \ .$$

We define an integer z by

(39.2) $$z = -(Bx + D) \ , \quad \text{so that} \quad x = -\frac{1}{B}(z + D) \ .$$

Clearly $z \neq 0$, for otherwise $x = -D/B$ and from (39.1) we would have $\dfrac{AD^2}{B^2} - \dfrac{CD}{B} + E = 0$, contradicting $F \neq 0$. From (39.1) and

(39.2) we have

$$B^2 zy = A(z+D)^2 - BC(z+D) + B^2 E ,$$

that is

$$z(B^2 y - Az - (2AD-BC)) = F ,$$

so that z is a divisor of F .

Thus the total number of possibilities for z is $\leq 2d(|F|)$.

For each such z there is at most one possibility for x , namely, $x = -\frac{1}{B}(z + D)$ if this is an integer. As (39.1) implies that each x determines at most one y , the total number of pairs (x,y) is $\leq 2d(|F|)$.

40. Evaluate $\displaystyle\sum_{k=1}^{n} \frac{k}{k^4 + k^2 + 1}$.

Solution: We have for $k \geq 1$

$$k^4 + k^2 + 1 = (k^4 + 2k^2 + 1) - k^2 = (k^2 - k + 1)(k^2 + k + 1)$$

and

$$\frac{2k}{k^4 + k^2 + 1} = \frac{1}{k^2 - k + 1} - \frac{1}{k^2 + k + 1}$$

$$= f(k-1) - f(k) ,$$

where

$$f(x) = \frac{1}{x^2 + x + 1} ,$$

so that

$$\sum_{k=1}^{n} \frac{k}{k^4 + k^2 + 1} = \frac{1}{2} \sum_{k=1}^{n} (f(k-1) - f(k))$$

$$= \frac{1}{2} (f(0) - f(n))$$

$$= \frac{1}{2} \left[1 - \frac{1}{n^2 + n + 1} \right]$$

$$= \frac{1}{2} \frac{n^2 + n}{n^2 + n + 1} \quad .$$

41. Let $P_m = P_m(n)$ denote the sum of all positive products of m different integers chosen from the set $\{1,2,\ldots,n\}$. Find formulae for $P_2(n)$ and $P_3(n)$.

Solution: We will show that

$$P_2 = \frac{1}{24} n(n+1)(n-1)(3n+2) \quad , \quad P_3 = \frac{1}{48} n^2(n+1)^2(n-1)(n-2) \quad .$$

We begin by considering

$$(1-x)(1-2x)\ldots(1-nx) = 1 - P_1 x + P_2 x^2 - P_3 x^3 + \ldots + (-1)^n P_n x^n \quad ,$$

so that, with $P_0 = 1$ and x sufficiently small,

$$\ell n \left(\sum_{r=0}^{n} (-1)^r P_r x^r \right) = \sum_{k=1}^{n} \ell n(1 - kx)$$

$$= - \sum_{k=1}^{n} \sum_{\ell=1}^{\infty} \frac{k^\ell x^\ell}{\ell}$$

$$= - \sum_{\ell=1}^{\infty} \frac{x^\ell}{\ell} \left(\sum_{k=1}^{n} k^\ell \right) \quad .$$

Hence we obtain

$$\sum_{r=0}^{n} (-1)^r P_r x^r = \exp \left(- \sum_{\ell=1}^{\infty} \frac{x^\ell}{\ell} \left(\sum_{k=1}^{n} k^\ell \right) \right) \quad ,$$

that is

$$(41.1) \qquad \sum_{r=0}^{n} (-1)^r P_r x^r = \sum_{h=0}^{\infty} \frac{1}{h!} \left[-\sum_{\ell=1}^{\infty} \frac{x^\ell}{\ell} \left(\sum_{k=1}^{n} k^\ell \right) \right]^h \ ,$$

and so, for $r = 0,1,2,\ldots,n$,

$$P_r = (-1)^r \text{ coefficient of } x^r \text{ on the right of } (41.1) \ .$$

Thus we have

$$P_2 = \text{coeff. of } x^2 \text{ in } \sum_{h=0}^{2} \frac{(-1)^h}{h!} \left[\sum_{\ell=1}^{\infty} \frac{x^\ell}{\ell} \left(\sum_{k=1}^{n} k^\ell \right) \right]^h$$

$$= -\frac{1}{2} \sum_{k=1}^{n} k^2 + \frac{1}{2!} \left(\sum_{k=1}^{n} k \right)^2$$

$$= -\frac{1}{2} \frac{n(n+1)(2n+1)}{6} + \frac{1}{2} \left(\frac{n(n+1)}{2} \right)^2$$

$$= \frac{1}{24} n(n+1)(n-1)(3n+2) \ ,$$

and

$$P_3 = \text{coeff. of } x^3 \text{ in } \sum_{h=0}^{3} \frac{(-1)^{h-1}}{h!} \left[\sum_{\ell=1}^{\infty} \frac{x^\ell}{\ell} \left(\sum_{k=1}^{n} k^\ell \right) \right]^h$$

$$= +\frac{1}{3} \left(\sum_{k=1}^{n} k^3 \right) - \frac{1}{2!} \left(\sum_{k=1}^{n} k \right) \left(\sum_{k=1}^{n} k^2 \right) + \frac{1}{3!} \left(\sum_{k=1}^{n} k \right)^3$$

$$= \frac{1}{12} n^2(n+1)^2 - \frac{1}{24} n^2(n+1)^2(2n+1) + \frac{1}{48} n^3(n+1)^3$$

$$= \frac{1}{48} n^2(n+1)^2(n-1)(n-2) \ .$$

42. For $a > b > 0$, evaluate the integral

$$(42.0) \qquad \int_0^{\infty} \frac{e^{ax} - e^{bx}}{x(e^{ax}+1)(e^{bx}+1)} \, dx \ .$$

Solution: For any constant C , the function

$$f(x) = \frac{e^x}{e^x + 1} + C \ , \quad x \geq 0 \ ,$$

is such that

$$f(ax) - f(bx) = \frac{e^{ax} - e^{bx}}{(e^{ax}+1)(e^{bx}+1)} \ .$$

For t > 0 we have

$$J(t) = \int_0^t \frac{e^{ax} - e^{bx}}{x(e^{ax}+1)(e^{bx}+1)} \, dx = \int_0^t \frac{f(ax) - f(bx)}{x} \, dx$$

$$= \int_0^t \frac{f(ax)}{x} \, dx - \int_0^t \frac{f(bx)}{x} \, dx \ ,$$

provided both the latter integrals exist.

This is guaranteed if $\lim\limits_{y \to 0} \frac{f(y)}{y}$ exists, which holds if and only if $C = -\frac{1}{2}$. With the choice $C = -\frac{1}{2}$, we have

$$J(t) = \int_0^{at} \frac{f(y)}{y} \, dy - \int_0^{bt} \frac{f(y)}{y} \, dy$$

$$= \int_{bt}^{at} \frac{f(y)}{y} \, dy \ .$$

Now $\lim\limits_{x \to \infty} f(x) = \frac{1}{2}$ so that given any $\varepsilon > 0$ there exists a positive real number $x_0 = x_0(\varepsilon)$ such that

$$x > x_0 \implies \frac{1}{2} - \varepsilon < f(x) < \frac{1}{2} + \varepsilon \ .$$

If $t > x_0/b$, then $at > bt > x_0$, and so we have

$$(\frac{1}{2} - \varepsilon) \, \ell n \frac{a}{b} = \int_{bt}^{at(\frac{1}{2} - \varepsilon)} \frac{dy}{y}$$

$$< J(t)$$

$$< \int_{bt}^{at(\frac{1}{2} + \varepsilon)} \frac{dy}{y}$$

$$= (\frac{1}{2} + \varepsilon) \, \ell n \frac{a}{b} \ .$$

Since ε is arbitrary we obtain

$$\lim_{t \to \infty} J(t) = \frac{1}{2} \ell n \frac{a}{b} \ ,$$

that is

$$\int_0^{\infty} \frac{e^{ax} - e^{bx}}{x(e^{ax}+1)(e^{bx}+1)} \, dx = \frac{1}{2} \ell n \frac{a}{b} \ .$$

43. For integers $n \geq 1$, determine the sum of n terms of the series

(43.0) $\dfrac{2n}{2n-1} + \dfrac{2n(2n-2)}{(2n-1)(2n-3)} + \dfrac{2n(2n-2)(2n-4)}{(2n-1)(2n-3)(2n-5)} + \cdots \ .$

Solution: Let S_n denote the sum of n terms of the given series (43.0). We have

$$S_1 = \frac{2}{1} = 2 \ ,$$

$$S_2 = \frac{4}{3} + \frac{4 \cdot 2}{3 \cdot 1} = \frac{4}{3} + \frac{8}{3} = \frac{12}{3} = 4 \ ,$$

$$S_3 = \frac{6}{5} + \frac{6 \cdot 4}{5 \cdot 3} + \frac{6 \cdot 4 \cdot 2}{5 \cdot 3 \cdot 1} = \frac{18+24+48}{15} = \frac{90}{15} = 6 \ .$$

These values suggest the conjecture $S_n = 2n$ for all positive integers n. First, as $S_1 = 2$, the conjecture holds for $n = 1$. Assume that $S_n = 2n$ holds for $n = m$.

Then we have

$$S_{m+1} = \frac{2m+2}{2m+1} + \frac{2m+2}{2m+1}\left[\frac{2m}{2m-1} + \frac{2m(2m-2)}{(2m-1)(2m-3)} + \ldots \text{(m terms)}\right]$$

$$= \frac{2m+2}{2m+1} + \frac{2m+2}{2m+1}\, S_m$$

$$= \frac{2m+2}{2m+1} + \frac{2m+2}{2m+1}\cdot 2m$$

$$= \frac{2m+2}{2m+1}\,(1 + 2m)$$

$$= 2m+2 ,$$

showing that $S_n = 2n$ is true for $n = m + 1$. Hence, by the principle of mathematical induction, $S_n = 2n$ is true for all positive integers n.

44. Let m be a fixed positive integer and let $z_1, z_2, \ldots, z_k$ be k (≥ 1) complex numbers such that

(44.0) $$z_1^s + z_2^s + \ldots + z_k^s = 0 ,$$

for all $s = m, m+1, m+2, \ldots, m+k-1$. Must $z_i = 0$ for $i = 1, 2, \ldots, k$?

Solution: The answer is yes. To see this, let $z_1, z_2, \ldots, z_k$ be the roots of the equation

$$z^k + a_{k-1}z^{k-1} + a_{k-2}z^{k-2} + \ldots + a_1 z + a_0 = 0 .$$

We will show that $a_0 = 0$. Suppose $a_0 \neq 0$. Clearly $z_1, z_2, \ldots, z_k$ are also roots of

$$z^{m+k-1} + a_{k-1}z^{m+k-2} + \ldots + a_1 z^m + a_0 z^{m-1} = 0 .$$

Hence for $i = 1, 2, \ldots, k$ we have

(44.1) $z_i^{m+k-1} + a_{k-1} z_i^{m+k-2} + \ldots + a_1 z_i^m + a_0 z_i^{m-1} = 0$.

Adding the equations in (44.1) and appealing to (44.0) we obtain

$$a_0 \sum_{i=1}^{k} z_i^{m-1} = 0 .$$

As $a_0 \neq 0$ we have

$$\sum_{i=1}^{k} z_i^{m-1} = 0 .$$

Clearly $z_1, z_2, \ldots, z_k$ are roots of

(44.2) $z^{m+k-2} + a_{k-1} z^{m+k-3} + \ldots + a_1 z^{m-1} + a_0 z^{m-2} = 0$.

Taking $z = z_i$, $i = 1, 2, \ldots, k$, in (44.2) and adding the equations we obtain as before

$$\sum_{i=1}^{k} z_i^{m-2} = 0 .$$

Repeating the argument we eventually obtain

$$\sum_{i=1}^{k} z_i = 0 ,$$

and one more application then gives $a_0 k = 0$, which is impossible. Hence we must have $a_0 = 0$, that is

$$(-1)^k z_1 \ldots z_k = 0 ,$$

and so at least one of the z_i is 0 , say $z_k = 0$. The argument can then be applied to $z_1, \ldots, z_{k-1}$ to prove that at least one of these is 0 , say $z_{k-1} = 0$. Continuing in this way we obtain

$$z_1 = z_2 = \ldots = z_k = 0 .$$

45. Let $A_n = (a_{ij})$ be the $n \times n$ matrix where

$$a_{ij} = \begin{cases} x, & \text{if } i = j, \\ 1, & \text{if } |i-j| = 1, \\ 0, & \text{otherwise,} \end{cases}$$

where $x > 2$. Evaluate $D_n = \det A_n$.

Solution: Expanding D_n by the first row, we obtain

$$D_n = x D_{n-1} - D_{n-2},$$

so that

(45.1) $D_n - x D_{n-1} + D_{n-2} = 0$.

The auxiliary equation for this difference equation is

$$t^2 - xt + 1 = 0,$$

which has the distinct real solutions

$$t = \frac{x \pm \sqrt{x^2 - 4}}{2},$$

as $x > 2$. Thus the solution of the difference equation (45.1) is given by

$$D_n = A\left(\tfrac{1}{2}(x + \sqrt{x^2 - 4})\right)^n + B\left(\tfrac{1}{2}(x - \sqrt{x^2 - 4})\right)^n,$$

for some constants A and B.

We now set for convenience

$$a = \tfrac{1}{2}(x + \sqrt{x^2 - 4}),$$

so that, as $\tfrac{1}{2}(x + \sqrt{x^2 - 4}) \cdot \tfrac{1}{2}(x - \sqrt{x^2 - 4}) = 1$,

$$\frac{1}{a} = \tfrac{1}{2}(x - \sqrt{x^2 - 4}),$$

which gives

$$D_n = Aa^n + Ba^{-n} , \quad n = 1,2,3,\ldots .$$

Now

$$D_1 = x = a + \frac{1}{a} = Aa + \frac{B}{a}$$

and

$$D_2 = x^2 - 1 = \left(a + \frac{1}{a}\right)^2 - 1 = a^2 + \frac{1}{a^2} + 1 = Aa^2 + \frac{B}{a^2}$$

so that

(45.2)
$$\begin{cases} a^2 A + B = a^2 + 1 , \\ a^4 A + B = a^4 + a^2 + 1 . \end{cases}$$

Solving (45.2) for A and B yields

$$A = \frac{a^2}{a^2 - 1} , \quad B = \frac{-1}{a^2 - 1} ,$$

so that

$$D_n = \frac{a^2}{a^2-1} a^n - \frac{1}{a^2-1} \frac{1}{a^n} ,$$

that is

$$D_n = \frac{a^{2n+2} - 1}{a^n(a^2-1)} , \quad n = 1,2,3,\ldots .$$

46. Determine a necessary and sufficient condition for the equations

(46.0)
$$\begin{cases} x + y + z = A , \\ x^2 + y^2 + z^2 = B , \\ x^3 + y^3 + z^3 = C , \end{cases}$$

to have a solution with at least one of x,y,z equal to zero.

Solution: Let x, y, z be a solution of (46.0). Then, from the identity

$$(x+y+z)^2 = x^2 + y^2 + z^2 + 2(xy+yz+zx) ,$$

and (46.0), we deduce that

(46.1) $$xy + yz + zx = \frac{1}{2}(A^2 - B) .$$

Next, from the identity

$$x^3 + y^3 + z^3 - 3xyz = (x + y + z)(x^2 + y^2 + z^2 - (xy+yz+zx)) ,$$

we obtain using (46.1)

$$C - 3xyz = A(B - \frac{1}{2}(A^2-B)) = \frac{3}{2}AB - \frac{1}{2}A^3 ,$$

so that

(46.2) $$3xyz = \frac{1}{2}A^3 - \frac{3}{2}AB + C .$$

Hence a solution (x,y,z) of (46.0) has at least one of x,y,z zero if and only if $xyz = 0$, that is, by (46.2), if and only if the condition $A^3 - 3AB + 2C = 0$ holds.

47. Let S be a set of k distinct integers chosen from $1,2,3,\ldots,10^n-1$, where n is a positive integer. Prove that if

(47.0) $$n < \ell n\left[\frac{(2^k-1)}{k} + \frac{(k+1)}{2}\right] \Big/ \ell n \ 10 ,$$

it is possible to find 2 disjoint subsets of S whose members have the same sum.

Solution: The integers in S are all $\leq 10^n-1$. Hence the sum of the integers in any subset of S is

$$\leq (10^n-k) + \ldots + (10^n-2) + (10^n - 1) = k \cdot 10^n - \frac{1}{2}k(k+1) .$$

The number of non-empty subsets of S is 2^k-1 . From (47.0) we have

$$2^k - 1 > k \cdot 10^n - \frac{1}{2}k(k+1) \ ,$$

and so, by Dirichlet's box principle, there must exist at least two different subsets of S , say S_1 and S_2 , which have the same sum. If S_1 and S_2 are disjoint the problem is solved.

If not, removal of the common elements from S_1 and S_2 yields two new subsets S_1' and S_2' with the required property.

48. Let n be a positive integer. Is it possible for $6n$ distinct straight lines in the Euclidean plane to be situated so as to have at least $6n^2-3n$ points where exactly three of these lines intersect and at least $6n+1$ points where exactly two of these lines intersect?

Solution: Any two distinct lines in the plane meet in at most one point. There are altogether $\binom{6n}{2} = 3n(6n-1)$ pairs of lines. A triple intersection accounts for 3 of these pairs of lines, and a simple intersection accounts for one pair. As

$$(6n^2-3n)3 + (6n+1)1$$
$$= 18n^2 - 3n + 1$$
$$> 3n(6n - 1)$$

the configuration is impossible.

49. Let S be a set with n (≥ 1) elements. Determine an explicit formula for the number $A(n)$ of subsets of S whose cardinality is a multiple of 3 .

<u>Solution:</u> The number of subsets of S containing 3ℓ elements is

$\binom{n}{3\ell}$, $\ell = 0,1,2,\ldots,[n/3]$. Thus, we have

(49.1)
$$A(n) = \sum_{\ell=0}^{[n/3]} \binom{n}{3\ell} = \sum_{\substack{k=0 \\ k\equiv 0 \ (\text{mod } 3)}}^{n} \binom{n}{k} .$$

Let $w = \frac{1}{2}(-1 + i\sqrt{3})$ so that

$$w^2 = \frac{1}{2}(-1 - i\sqrt{3}) , \quad w^3 = 1 ,$$

and, for $r = 0,1,2$, define

$$S_r = \sum_{\substack{k=0 \\ k\equiv r \ (\text{mod } 3)}}^{n} \binom{n}{k} .$$

Then, by the binomial theorem, we have

(49.2)
$$2^n = (1+1)^n = \sum_{k=0}^{n} \binom{n}{k} = S_0 + S_1 + S_2 ,$$

(49.3)
$$(1 + w)^n = \sum_{k=0}^{n} \binom{n}{k} w^k = S_0 + wS_1 + w^2 S_2 ,$$

(49.4)
$$(1 + w^2)^n = \sum_{k=0}^{n} \binom{n}{k} w^{2k} = S_0 + w^2 S_1 + wS_2 .$$

Adding (49.2), (49.3), (49.4), we obtain, as $1 + w + w^2 = 0$,

$$2^n + (1+w)^n + (1+w^2)^n = 3S_0 = 3A(n) ,$$

so that

$$A(n) = \frac{1}{3}(2^n + (-w^2)^n + (-w)^n) .$$

Hence we have

$$A(n) = \begin{cases} \dfrac{1}{3}(2^n + 2(-1)^n) , & \text{if } n \equiv 0 \ (\text{mod } 3) , \\[2mm] \dfrac{1}{3}(2^n - (-1)^n) , & \text{if } n \not\equiv 0 \ (\text{mod } 3) . \end{cases}$$

50. For each integer $n \geq 1$, prove that there is a polynomial $p_n(x)$ with integral coefficients such that

$$x^{4n}(1-x)^{4n} = (1+x^2)p_n(x) + (-1)^n 4^n .$$

Define the rational number a_n by

(50.0) $$a_n = \frac{(-1)^{n-1}}{4^{n-1}} \int_0^1 p_n(x) \, dx , \quad n = 1,2,\ldots .$$

Prove that a_n satisfies the inequality

$$\left| \pi - a_n \right| < \frac{1}{4^{5n-1}} , \quad n = 1,2,\ldots .$$

<u>Solution</u> (due to L. Smith): Let Z denote the domain of rational integers and $Z[i] = \{a+bi: a,b \in Z\}$ the domain of gaussian integers. For $n = 1,2,3,\ldots$ set

(50.1) $$q_n(x) = x^{4n}(1-x)^{4n} - (-1)^n 4^n ,$$

so $q_n(x) \in Z[x]$. As $q_n(\pm i) = 0$, $q_n(x)$ is divisible by $x+i$ and $x-i$ in $Z[i][x]$, and so $p_n(x) = q_n(x)/(x^2+1) \in Z[i][x]$. However $p_n(x) \in R[x]$, and as $R[x] \cap Z[i][x] = Z[x]$, we have $p_n(x) \in Z[x]$. This proves the first part of the question.

For the second part, we note that

$$\frac{x^{4n}(1-x)^{4n}}{1+x^2} = p_n(x) + \frac{(-1)^n 4^n}{1+x^2} ,$$

so that

$$\int_0^1 \frac{x^{4n}(1-x)^{4n}}{1+x^2} \, dx = \int_0^1 p_n(x) \, dx + (-1)^n 4^n \int_0^1 \frac{dx}{1+x^2} .$$

Now, as $\int_0^1 \frac{dx}{1+x^2} = \frac{\pi}{4}$, we have using (50.0)

$$\left|\pi - a_n\right| = \frac{1}{4^{n-1}} \int_0^1 \frac{x^{4n}(1-x)^{4n}}{1+x^2}\, dx \ .$$

Now

$$\frac{x^{4n}(1-x)^{4n}}{1+x^2} \leq x^{4n}(1-x)^{4n} \leq \frac{1}{4^{4n}} \ ,$$

as $x(1-x) \leq \frac{1}{4}$, and thus we have

$$\left|\pi - a_n\right| < \frac{1}{4^{5n-1}} \ ,$$

completing the second part of the question.

51. In last year's boxing contest, each of the 23 boxers from the blue team fought exactly one of the 23 boxers from the green team, in accordance with the contest regulation that opponents may only fight if the absolute difference of their weights is less than one kilogram.

Assuming that this year the members of both teams remain the same as last year and that their weights are unchanged, show that the contest regulation is satisfied if the lightest member of the blue team fights the lightest member of the green team, the next lightest member of the blue team fights the next lightest member of the green team, and so on.

Solution: More generally we consider teams, each containing n members, such that the absolute difference of weights of opponents last year was less than d kilograms.

Let $B_1, B_2, \ldots, B_n$ denote the members of the blue team with weights

$$b_1 \leq b_2 \leq \ldots \leq b_n \ ,$$

and let $G_1, G_2, \ldots, G_n$ denote the members of the green team with weights

$$g_1 \leq g_2 \leq \ldots \leq g_n .$$

For each r with $1 \leq r \leq n$, we consider this year's opponents B_r and G_r . We show that $|b_r - g_r| \leq d$. We treat only the case $b_r \geq g_r$ as the case $b_r \leq g_r$ is similar. If there exists s with $r < s \leq n$ and t with $1 \leq t \leq r$ such that B_s fought G_t last year, then $b_r - g_r \leq b_s - g_t \leq d$. If not, then every boxer B_s with $r < s \leq n$ was paired with an opponent G_t with $r < t \leq n$ last year, and thus B_r must have been paired with some G_u with $1 \leq u \leq r$ last year. Thus we have

$$b_r - g_r \leq b_r - g_u \leq d .$$

This completes the proof.

52. Let S be the set of all composite positive odd integers less than 79 .

(a) Show that S may be written as the union of three (not necessarily disjoint) arithmetic progressions.

(b) Show that S cannot be written as the union of two arithmetic progressions.

Solution: (a) Each member of S can be written in the form

$$(2r + 1)(2r + 2s + 1) ,$$

for suitable integers $r \geq 1$ and $s \geq 0$, and so belongs to the arithmetic progression with first term $(2r+1)^2$ and common difference $2(2r+1)$. Taking $r = 1, 2, 3$ we define arithmetic progressions A_1, A_2, A_3 as follows:

$$A_1 = \{9,15,21,27,33,39,45,51,57,63,69,75\}\ ,$$

$$A_2 = \{25,35,45,55,65,75\}\ ,$$

$$A_3 = \{49,63,77\}\ .$$

It is easily checked that

$$S = A_1 \cup A_2 \cup A_3\ .$$

(b)   Suppose that

$$S = A \cup B\ ,$$

where

$$A = \{a\ ,\ a+b\ ,\ \ldots\ ,\ a+(m-1)b\}$$

and

$$B = \{c\ ,\ c+d\ ,\ \ldots\ ,\ c+(n-1)d\}\ .$$

Without loss of generality we may take $a = 9$. Then we have either $a + b = 15$ or $c = 15$. In the former case $A = A_1$ and so $c = 25$, $c + d = 35$ giving $B = A_2$. This is impossible as 49 is neither in A nor B . In the latter case either $a + b = 21$ or $c + d = 21$. If $a + b = 21$ we have $A = \{9,21,33,45,57,69\}$ and so $c + d = 27$ giving $B = \{15,27,39,51,63,75\}$. This is impossible as 49 belongs neither to A nor B . If $c + d = 21$ we have $B = A_1 - \{9\}$ so $a + b = 25$ giving $A = \{9,25,41,\ldots\}$ which is impossible as 41 is prime.

53.   For $b > 0$, prove that

$$\left| \int_0^b \frac{\sin x}{x}\, dx - \frac{\pi}{2} \right| < \frac{1}{b}\ ,$$

by first showing that

$$\int_0^b \frac{\sin x}{x}\, dx = \int_0^\infty \left(\int_0^b e^{-ux} \sin x\, dx \right) du\ .$$

Solution: We begin by showing that for b > 0

(53.1) $$\int_0^b \frac{\sin x}{x} dx = \lim_{y \to \infty} \int_0^b (1 - e^{-xy}) \frac{\sin x}{x} dx .$$

We have for y > 0

$$\left| \int_0^b \frac{\sin x}{x} dx - \int_0^b (1 - e^{-xy}) \frac{\sin x}{x} dx \right|$$

$$= \left| \int_0^b e^{-xy} \frac{\sin x}{x} dx \right|$$

$$\leq \max_{0 \leq x \leq b} \left| \frac{\sin x}{x} \right| \int_0^b e^{-xy} dx$$

$$= M(b) \left. \frac{e^{-xy}}{-y} \right|_0^b$$

$$= M(b) \frac{(1 - e^{-by})}{y}$$

$$\leq \frac{M(b)}{y} .$$

Letting $y \to \infty$ we obtain (53.1).

Next we have

$$\int_0^b (1 - e^{-xy}) \frac{\sin x}{x} dx$$

$$= \int_0^b \left[\int_0^y x e^{-xu} du \right] \frac{\sin x}{x} dx$$

$$= \int_0^y \left[\int_0^b e^{-ux} \sin x \, dx \right] du .$$

Letting $y \to \infty$ we obtain

$$\int_0^b \frac{\sin x}{x} dx = \int_0^\infty \left[\int_0^b e^{-ux} \sin x \, dx \right] du$$

$$= \int_0^\infty \frac{(1 - e^{-bu}(u \sin b + \cos b))}{1 + u^2} \, du$$

$$= \frac{\pi}{2} - \int_0^\infty \frac{e^{-bu}(u \sin b + \cos b)}{1 + u^2} \, du \ .$$

Hence we have

$$\left| \int_0^b \frac{\sin x}{x} \, dx - \frac{\pi}{2} \right|$$

$$= \left| \int_0^\infty \frac{e^{-bu}(u \sin b + \cos b)}{1 + u^2} \, du \right|$$

$$\leq \int_0^\infty e^{-bu} \frac{\sqrt{1+u^2}}{1+u^2} \, du$$

$$\leq \int_0^\infty e^{-bu} du$$

$$= \frac{1}{b} \ ,$$

as required.

54. Let $a_1, a_2, \ldots, a_{44}$ be 44 natural numbers such that

$$0 < a_1 < a_2 < \ldots < a_{44} \leq 125 \ .$$

Prove that at least one of the 43 differences $d_j = a_{j+1} - a_j$ occurs at least 10 times.

Solution: We have

$$\sum_{j=1}^{43} d_j = \sum_{j=1}^{43} (a_{j+1} - a_j) = a_{44} - a_1 \leq 125 - 1 = 124 \ .$$

If each difference d_j occurs at most 9 times then

$$\sum_{j=1}^{43} d_j \geq 9(1+2+3+4) + 7(5) = 125 .$$

This is clearly a contradiction so at least one difference must occur at least 10 times.

55. Show that for every natural number n there exists a prime p such that $p = a^2 + b^2$, where a and b are natural numbers both greater than n . (You may appeal to the following two theorems:

(A) If p is a prime of the form 4t+1 then there exist integers a and b such that $p = a^2 + b^2$.

(B) If r and s are natural numbers such that GCD(r,s) = 1 , there exist infinitely many primes of the form rk+s , where k is a natural number.)

Solution: Let n be a natural number. By (B) there exists a prime q > n of the form 4t+3 . Set

$$m = 2(1^2+q)(2^2+q)\ldots(n^2+q) .$$

Clearly we have

$$GCD(m,q) = 1 .$$

Hence, by (B), there exists a natural number k such that the number

$$p = m^2 k - q$$

is a prime. Clearly p is of the form 4u+1 . Hence, by (A), there exist natural numbers a and b such that

$$p = a^2 + b^2 .$$

Without loss of generality we may assume that a < b . Suppose now that a ≤ n . Then we have

$$b^2 = p - a^2 = m^2 k - q - a^2$$

$$= 4k \prod_{r=1}^{n} (r^2+q)^2 - (a^2+q)$$

$$= (a^2+q) \left[4(a^2+q) \prod_{\substack{r=1 \\ r \neq a}}^{n} (r^2+q)^2 - 1 \right] ,$$

where the factors on the right hand side of the equality are coprime. Consequently they must be squares, but this is impossible as the second factor is of the form $4v-1$. Thus we must have $b > a > n$.

56. Let $a_1, a_2, \ldots, a_n$ be n (≥ 1) integers such that
(i) $0 < a_1 < a_2 < \ldots < a_n$,
(ii) all the differences $a_i - a_j$ $(1 \leq j < i \leq n)$ are distinct,
(iii) $a_i \equiv a \pmod{b}$ $(1 \leq i \leq n)$, where a and b are positive
      integers such that $1 \leq a \leq b-1$.
Prove that

$$\sum_{r=1}^{n} a_r \geq \frac{b}{6} n^3 + (a - \frac{b}{6}) n .$$

Solution: Let r be an integer such that $2 \leq r \leq n$. For $1 \leq j < i \leq r$ there are $\binom{r}{2} = \frac{1}{2} r(r-1)$ distinct differences $a_i - a_j$, and these are all divisible by b . Thus the largest difference among these, namely $a_r - a_1$, must be at least $\frac{b}{2} r(r-1)$, that is

$$a_r - a_1 \geq \frac{b}{2} r(r-1) , \quad 2 \leq r \leq n ,$$

and so

$$a_r \geq a_1 + \frac{b}{2} r(r-1) , \quad 2 \leq r \leq n .$$

As $a_1 \equiv a \pmod{b}$ there is an integer t such that $a_1 = a + bt$. If $t \leq -1$ then $a_1 = a + bt \leq a - b \leq -1$, which is impossible as $a_1 > 0$. Hence we have $t \geq 0$ and so $a_1 \geq a$, giving

$$(56.1) \qquad a_r \geq a + \frac{b}{2} r(r-1), \quad 2 \leq r \leq n.$$

The inequality (56.1) clearly holds for $r = 1$. Thus we have

$$\sum_{r=1}^{n} a_r \geq an + \frac{b}{2} \sum_{r=1}^{n} r(r-1) = an + \frac{b}{2} \cdot \frac{n(n^2-1)}{3},$$

so that

$$\sum_{r=1}^{n} a_r \geq \frac{b}{6} n^3 + (a - \frac{b}{6})n.$$

57. Let $A_n = (a_{ij})$ be the $n \times n$ matrix given by

$$a_{ij} = \begin{cases} 2\cos t, & \text{if } i = j, \\ 1, & \text{if } |i - j| = 1, \\ 0, & \text{otherwise}, \end{cases}$$

where $-\pi < t < \pi$. Evaluate $D_n = \det A_n$.

Solution: Expanding $D_n = \det A_n$ by the first row, we obtain the recurrence relation

$$(57.1) \qquad D_n = 2\cos t \, D_{n-1} - D_{n-2}, \quad n \geq 2.$$

We now consider two cases according as $t \neq 0$ or $t = 0$. For $t \neq 0$ the values of D_1 and D_2 may be obtained by direct calculation as follows:

$$D_1 = 2\cos t = \frac{\sin 2t}{\sin t},$$

and

$$D_2 = 4 \cos^2 t - 1$$

$$= \frac{4 \sin t \cos^2 t - \sin t}{\sin t}$$

$$= \frac{2 \sin t \cos^2 t + \sin t (2 \cos^2 t - 1)}{\sin t}$$

$$= \frac{\sin 2t \cos t + \sin t \cos 2t}{\sin t}$$

$$= \frac{\sin (2t + t)}{\sin t}$$

$$= \frac{\sin 3t}{\sin t} .$$

These values suggest that

(57.2) $$D_n = \frac{\sin (n+1)t}{\sin t} , \quad n = 1,2,3,\ldots .$$

In order to prove (57.2), assume that (57.2) holds for $n = 1,2,\ldots,k-1$.
Then we have, by the recurrence relation (57.1),

$$D_k = 2 \cos t \, D_{k-1} - D_{k-2}$$

$$= 2 \cos t \, \frac{\sin kt}{\sin t} - \frac{\sin (k-1)t}{\sin t}$$

$$= \frac{2 \sin kt \cos t - \sin (k-1)t}{\sin t}$$

$$= \frac{\sin (k+1)t}{\sin t} .$$

Thus (57.2) holds for all n by the principle of mathematical induction.

For $t = 0$ we have

$$D_1 = 2 , \quad D_2 = 3 , \quad D_3 = 4 ,$$

which suggests that $D_n = n + 1$, $n \geq 1$. This can also be proved
by mathematical induction. We note that

$$\lim_{t \to 0} \frac{\sin (n+1)t}{\sin t} = n + 1 .$$

Remark: The auxiliary equation for the difference equation (57.1) is

$$x^2 - 2(\cos t)x + 1 = 0 ,$$

which has the roots

$$x = \begin{cases} \exp(it) , \exp(-it) , & t \neq 0 , \\ 1 \text{ (repeated)} & , t = 0 , \end{cases}$$

giving

$$D_n = \begin{cases} A_1\exp(nit) + B_1\exp(-nit) , & t \neq 0 , \\ A_2 + B_2 n & , t = 0 , \end{cases}$$

for complex constants A_1 and B_1 and real constants A_2 and B_2, which can be determined from the initial values $D_1 = 2 \cos t$, $D_2 = 4\cos^2 t - 1$, using DeMoivre's theorem in the case $t \neq 0$. One finds

$$A_1 = \frac{1}{2}(1 - i \cot t) , B_1 = \frac{1}{2}(1 + i \cot t) ,$$

$$A_2 = B_2 = 1 .$$

58. Let a and b be fixed positive integers. Find the general solution of the recurrence relation

(58.0) $\qquad x_{n+1} = x_n + a + \sqrt{b^2 + 4ax_n}$, $n = 0,1,2,\ldots$,

where $x_0 = 0$.

Solution: From (58.0) we have

$$b^2 + 4ax_{n+1} = b^2 + 4a(x_n + a + \sqrt{b^2+4ax_n})$$

$$= 4a^2 + 4a\sqrt{b^2+4ax_n} + (b^2+4ax_n)$$

$$= (2a + \sqrt{b^2 + 4ax_n})^2 \; ,$$

so that

(58.1) $\sqrt{b^2 + 4ax_{n+1}} = 2a + \sqrt{b^2 + 4ax_n} \; .$

Hence, by (58.0) and (58.1), we have

(58.2) $x_n = x_{n+1} + a - \sqrt{b^2 + 4ax_{n+1}} \; .$

Replacing n by n−1 in (58.2) , we get

(58.3) $x_{n-1} = x_n + a - \sqrt{b^2 + 4ax_n} \; .$

Adding (58.0) and (58.3) we obtain

$$x_{n+1} + x_{n-1} = 2x_n + 2a \; ,$$

and hence

(58.4) $x_{n+1} - 2x_n + x_{n-1} = 2a \; .$

Setting $y_{n-1} = x_n - x_{n-1}$ in (58.4), we obtain

(58.5) $y_n - y_{n-1} = 2a \; .$

Adding (58.5) for n = 1,2,3,... yields (as $y_0 = x_1 - x_0 = a + b$)

$$y_n = 2an + (a + b) \; ,$$

and so $x_n = an^2 + bn \; .$

  59. Let a be a fixed real number satisfying $0 < a < \pi$, and set

(59.0) $I_r = \displaystyle\int_{-a}^{a} \frac{1 - r\cos u}{1 - 2r\cos u + r^2} \, du \; .$

Prove that

$$I_1 \; , \quad \lim_{r \to 1^+} I_r \; , \quad \lim_{r \to 1^-} I_r$$

all exist and are distinct.

Solution: We begin by calculating I_1 . We have

(59.1) $$I_1 = \int_{-a}^{a} \frac{1 - \cos u}{2 - 2 \cos u} \, du = \int_{-a}^{a} \frac{1}{2} \, du = a \ ,$$

where we have taken the value of the integrand to be $\frac{1}{2}$ when $u = 0$.

Now, for $r > 0$ and $r \neq 1$, we have

$$1 - 2r \cos u + r^2 > 2r - 2r \cos u = 2r(1 - \cos u) \geq 0 \ ,$$

so that the integrand of the integral in (59.0) is continuous on $[-a,a]$. We have

$$I_r = \int_{-a}^{a} \left\{ \frac{1}{2} + \frac{(1 - r^2)}{2(1 - 2r \cos u + r^2)} \right\} \, du$$

$$= a + \frac{1}{2}(1 - r^2) \int_{-a}^{a} \frac{du}{1 - 2r \cos u + r^2} \ ,$$

which gives

(59.2) $$I_r = a + \frac{(1 - r^2)}{2} J_r \ ,$$

where

(59.3) $$J_r = \int_{-a}^{a} \frac{du}{1 - 2r \cos u + r^2} \ .$$

Let $t = \tan \frac{u}{2}$ with $-a \leq u \leq a$ so that

$$\cos u = \frac{1 - t^2}{1 + t^2} \ , \quad du = \frac{2}{1 + t^2} \, dt \ .$$

Using the above transformation in (59.3), we obtain

(59.4)
$$J_r = \frac{2}{(1+r)^2} \int_{-t_1}^{t_1} \frac{dt}{t^2 + \left(\frac{1-r}{1+r}\right)^2} \,,$$

where

$$t_1 = \tan a/2 \,.$$

Evaluating the standard integral in (59.4), we deduce that

(59.5)
$$J_r = \frac{4}{|1-r^2|} \tan^{-1}\left(\left|\frac{1+r}{1-r}\right| \tan a/2\right) \,.$$

From (59.2) and (59.5) we get

$$I_r = a + 2\frac{(1-r^2)}{|1-r^2|} \tan^{-1}\left(\left|\frac{1+r}{1-r}\right| \tan a/2\right) \,.$$

Hence for $r > 1$ we have

$$I_r = a - 2\tan^{-1}\left(\left|\frac{r+1}{r-1}\right| \tan a/2\right) \,,$$

and for $0 < r < 1$ we have

$$I_r = a + 2\tan^{-1}\left(\left(\frac{1+r}{1-r}\right) \tan a/2\right) \,.$$

Taking limits we obtain

$$\lim_{r \to 1^+} I_r = a - 2 \cdot \frac{\pi}{2} = a - \pi \,,$$

$$\lim_{r \to 1^-} I_r = a + 2 \cdot \frac{\pi}{2} = a + \pi \,.$$

Thus the quantities I_1 , $\lim\limits_{r \to 1^+} I_r$, $\lim\limits_{r \to 1^-} I_r$ all exist and are all distinct.

60. Let I denote the class of all isosceles triangles. For $\Delta \varepsilon I$, let h_Δ denote the length of each of the two equal altitudes of Δ and k_Δ the length of the third altitude. Prove that there does not exist a function f of h_Δ such that

$$k_\Delta \leq f(h_\Delta) \ ,$$

for all $\Delta \varepsilon I$.

Solution: Let h be a fixed positive real number. For $t > 1$ let

$$(60.1) \qquad a = \frac{h(t + \frac{1}{t})^2}{4(t - \frac{1}{t})} \ , \quad b = \frac{h}{2}(t + \frac{1}{t}) \ .$$

As $b < 2a$ there exists an isosceles triangle $\Delta(t)$ with vertices A,B,C such that $|AB| = |AC| = a$, $|BC| = b$. It will follow that the choice (60.1) is such that

$$h_{\Delta(t)} = h \ , \quad k_{\Delta(t)} = \frac{h}{2}\left(\frac{t + \frac{1}{t}}{t - \frac{1}{t}} \right) \ .$$

Let P, Q, R be the feet of the perpendiculars from A to BC, B to CA, C to AB respectively. Then we have

$$h_{\Delta(t)}^2 = BQ^2 = CR^2 = a^2 \sin^2 A = a^2(1 - \cos^2 A) \ .$$

Applying the cosine law to $\Delta(t)$ we obtain

$$\cos A = \frac{2a^2 - b^2}{2a^2} \ .$$

Hence it follows that

$$(60.2) \qquad h_{\Delta(t)}^2 = a^2\left[1 - \left(\frac{2a^2 - b^2}{2a^2}\right)^2\right] = \frac{b^2}{4a^2}(4a^2 - b^2) \ .$$

Next, from (60.1), we see that

$$\frac{2a-b}{h} = \frac{(t + \frac{1}{t})}{t(t - \frac{1}{t})} \ , \quad \frac{2a+b}{h} = \frac{t(t + \frac{1}{t})}{(t - \frac{1}{t})} \ , \quad \frac{b^2}{4a^2} = \frac{(t - \frac{1}{t})^2}{(t + \frac{1}{t})^2} \ ,$$

so that

$$\frac{b^2}{4a^2}(4a^2 - b^2) = h^2 ,$$

and thus from (60.2) we have

$$h_{\Delta(t)} = h .$$

Applying Pythagoras' theorem in triangle ABP, we have

$$k^2_{\Delta(t)} = AP^2 = a^2 - \frac{b^2}{4} = \frac{h^2}{4}\frac{(t + \frac{1}{t})^2}{(t - \frac{1}{t})^2} ,$$

so that

$$k_{\Delta(t)} = \frac{h}{2}\left|\frac{t + \frac{1}{t}}{t - \frac{1}{t}}\right| .$$

Finally, suppose there exists a function $f = f(h_\Delta)$ such that

$$k_\Delta \leq f(h_\Delta) ,$$

for all $\Delta \in I$. Then, in particular, one sees that

$$k_{\Delta(t)} \leq f(h_{\Delta(t)}) , \quad t > 1 ,$$

which implies

$$\frac{h}{2}\left|\frac{t + \frac{1}{t}}{t - \frac{1}{t}}\right| \leq f(h) , \quad t > 1 ,$$

that is

(60.3) $$\frac{t + \frac{1}{t}}{t - \frac{1}{t}} \leq \frac{2\ f(h)}{h} , \quad t > 1 .$$

As the left side of (60.3) tends to infinity as $t \to +\infty$ while the
right side is fixed, we have obtained a contradiction, and therefore
no such function f can exist.

61. Find the minimum value of the expression

$$(61.0) \quad \left[x^2 + \frac{k^2}{x^2} \right] - 2 \left[(1+\cos t)x + \frac{k(1+\sin t)}{x} \right] + (3 + 2 \cos t + 2 \sin t),$$

for $x > 0$ and $0 \le t \le 2\pi$, where $k > \frac{3}{2} + \sqrt{2}$ is a fixed real number.

Solution: The expression given in (61.0) can be written in the form

$$(x - (1+\cos t))^2 + (\frac{k}{x} - (1+\sin t))^2,$$

which is the square of the distance between the point $(x, \frac{k}{x})$ $(x > 0)$ on the rectangular hyperbola $xy = k$ in the first quadrant, and the point $(1+\cos t, 1+\sin t)$ $(0 \le t \le 2\pi)$ on the circle centre $(1,1)$ with radius 1. The condition $k > \frac{3}{2} + \sqrt{2}$ ensures that the two curves are non-intersecting. Clearly the minimum distance between these two curves occurs for the point $(\sqrt{k}, \sqrt{k})$ on the hyperbola and the point $(1 + \frac{1}{\sqrt{2}}, 1 + \frac{1}{\sqrt{2}})$ on the circle. Hence the required minimum is

$$2(\sqrt{k} - (1 + \frac{1}{\sqrt{2}}))^2.$$

62. Let $\varepsilon > 0$. Around every point in the xy-plane with integral co-ordinates draw a circle of radius ε. Prove that every straight line through the origin must intersect an infinity of these circles.

Solution: Let L be a line through the origin with slope b. If b is rational, say $b = \frac{k}{\ell}$, where k and ℓ are integers satisfying $\ell \ge 1$ and $(k, \ell) = 1$, then L passes through the centres of the infinity of circles all of radius ε, centred at lattice points $(\ell t, kt)$, where t is an integer.

If b is irrational, then by Hurwitz's theorem there are infinitely many pairs of integers (m,n) with $n \neq 0$ and $GCD(m,n) = 1$ for which

$$\left| \frac{m}{n} - b \right| < \frac{1}{\sqrt{5}\,n^2} < \frac{1}{n^2} \; .$$

Choosing only those pairs (m,n) for which

$$n > \frac{1}{\varepsilon\sqrt{b^2+1}} \; ,$$

we see that there are infinitely many pairs (m,n) for which

$$\left| \frac{m}{n} - b \right| < \frac{\varepsilon\sqrt{b^2+1}}{n} \; ,$$

that is, for which

(62.1) $$\left| \frac{m - bn}{\sqrt{b^2+1}} \right| < \varepsilon \; .$$

Since the left side of (62.1) is the distance between the line L and the point (m,n), L intersects the infinity of circles all with radius ε centred at these lattice points.

<u>Second solution</u> (due to L. Smith): Let L be a line through the origin with slope b. The case when b is rational is treated as in the first solution.

When b is irrational, we construct an infinite sequence of lattice points whose distances from L are less than any given positive ε.

Pick any lattice point (x_1, y_1) with $d_1 = |y_1 - bx_1| < 1$. Clearly d_1 is positive as b is irrational. Set

$$a_1 = \left[\frac{1}{d_1} \right] \geq 1 \; ,$$

so that

(62.2)
$$\frac{1}{a_1+1} < d_1 \le \frac{1}{a_1} \quad .$$

Let (x_2, y_2) be the lattice point given by

$$(x_2, y_2) = \begin{cases} (a_1 x_1, a_1 y_1 - 1), & \text{if } y_1 > bx_1, \\ (a_1 x_1, a_1 y_1 + 1), & \text{if } y_1 < bx_1, \end{cases}$$

and set $d_2 = |y_2 - bx_2|$. Clearly, as $d_2 \ne 0$, we may set

(62.3)
$$a_2 = \left[\frac{1}{d_2}\right] \quad .$$

It is easy to see that $d_2 = 1 - a_1 d_1$, so that by (62.2) we have

(62.4)
$$d_2 < \frac{1}{a_1 + 1} \quad .$$

Thus, from (62.3) and (62.4), we obtain

$$a_1 < a_2 \quad .$$

Continuing this process we obtain an infinite sequence of lattice points $\{(x_k, y_k): k = 1, 2, \ldots\}$, whose vertical distances d_k from L satisfy

$$d_k < \frac{1}{a_{k-1} + 1}, \quad k \ge 2 ,$$

where $a_k = \left[\frac{1}{d_k}\right]$, $k \ge 1$. Furthermore $\{a_k: k = 1, 2, \ldots\}$ is a strictly increasing sequence of positive integers.

Finally choose a positive integer N such that $\frac{1}{N} < \varepsilon$. Then, for all $n \ge N + 1$, we have

$$d_n < \frac{1}{a_{n-1} + 1} < \frac{1}{a_N} \le \frac{1}{N} < \varepsilon ,$$

and the lattice points (x_n, y_n) $(n > N)$ are as required.

63. Let n be a positive integer. For $k = 0,1,2,\ldots,2n-2$ define

(63.0)
$$I_k = \int_0^\infty \frac{x^k}{x^{2n}+x^n+1} \, dx \quad .$$

Prove that $I_k \geq I_{n-1}$, $k = 0,1,2,\ldots,2n-2$.

Solution: For $k = 0,1,2,\ldots,2n-2$, (63.0) can be written

$$I_k = \lim_{\substack{a \to \infty \\ b \to 0^+}} \int_b^a \frac{x^k}{x^{2n}+x^n+1} \, dx \quad .$$

Applying the transformation $x = \frac{1}{y}$, we obtain

$$I_k = \lim_{\substack{a \to \infty \\ b \to 0^+}} \int_{1/a}^{1/b} \frac{y^{2n-k-2}}{y^{2n}+y^n+1} \, dy \quad ,$$

so that $I_k = I_{2n-k-2}$.

Now, using the arithmetic mean–geometric mean inequality, we have for $x \geq 0$,

(63.1)
$$\frac{x^k + x^{2n-k-2}}{2} \geq x^{n-1}, \quad k = 0,1,2,\ldots,2n-2 \quad .$$

As $x^{2n}+x^n+1 > 0$, we may divide (63.1) by $x^{2n}+x^n+1$, and integrate the resulting inequality to obtain

$$\frac{I_k + I_{2n-k-2}}{2} \geq I_{n-1} \quad ,$$

from which the desired inequality follows.

64. Let D be the region in Euclidean n-space consisting of all n-tuples $(x_1, x_2, \ldots, x_n)$ satisfying

$$x_1 \geq 0 \; , \quad x_2 \geq 0 \; , \quad \ldots \; , \quad x_n \geq 0 \; , \quad x_1 + x_2 + \ldots + x_n \leq 1 \; .$$

Evaluate the multiple integral

(64.0) $$\iint \ldots \int_D x_1^{k_1} x_2^{k_2} \ldots x_n^{k_n} (1-x_1-x_2-\ldots-x_n)^{k_{n+1}} dx_1 \ldots dx_n \; ,$$

where $k_1, \ldots, k_{n+1}$ are positive integers.

Solution (due to L. Smith): We begin by considering

$$I(r,s) = \int_0^a x^r (a-x)^s dx \; ,$$

where r and s are positive integers and $a > 0$.

Integrating $I(r,s)$ by parts, we obtain

$$I(r,s) = \frac{s}{r+1} I(r+1, s-1) \; ,$$

so that

$$I(r,s) = \frac{s}{r+1} \cdot \frac{s-1}{r+2} \ldots \frac{1}{r+s} I(r+s, 0) \; ,$$

that is

$$I(r,s) = \frac{r! s!}{(r+s)!} I(r+s, 0) \; .$$

As $I(k,0) = \dfrac{a^k}{k+1}$, for $k \geq 1$, we obtain

(64.1) $$I(r,s) = \frac{r! \, s! \, a^{r+s+1}}{(r+s+1)!} \; .$$

Now, applying (64.1) successively, we obtain

$$\iint_D \cdots \int x_1^{k_1} x_2^{k_2} \cdots x_n^{k_n} (1-x_1-x_2-\ldots-x_n)^{k_{n+1}} dx_1 \ldots dx_n$$

$$= \int_{x_1=0}^{1} x_1^{k_1} \int_{x_2=0}^{1-x_1} x_2^{k_2} \cdots \int_{x_{n-1}=0}^{1-x_1-x_2-\ldots-x_{n-2}} x_{n-1}^{k_{n-1}} \int_{x_n=0}^{1-x_1-\ldots-x_{n-1}} x_n^{k_n} ((1-x_1-\ldots-x_{n-1})-x_n)^{k_{n+1}} dx_n \ldots dx_1$$

$$= \frac{k_n! \, k_{n+1}!}{(k_n+k_{n+1}+1)!} \int_{x_1=0}^{1} x_1^{k_1} \int_{x_2=0}^{1-x_1} x_2^{k_2} \cdots \int_{x_{n-1}=0}^{1-x_1-\ldots-x_{n-2}} x_{n-1}^{k_{n-1}} (1-x_1-\ldots-x_{n-1})^{k_n+k_{n+1}+1} dx_{n-1} \ldots dx_1$$

$$= \frac{k_{n-1}! \, k_n! \, k_{n+1}!}{(k_{n-1}+k_n+k_{n+1}+2)!} \int_{x_1=0}^{1} x_1^{k_1} \int_{x_2=0}^{1-x_1} x_2^{k_2} \cdots \int_{x_{n-2}=0}^{1-x_1-\ldots-x_{n-3}} x_{n-2}^{k_{n-2}} (1-x_1-\ldots-x_{n-2})^{k_{n-1}+k_n+k_{n+1}+2} dx_{n-2} \ldots dx_1$$

$$= \ldots$$

$$= \frac{k_1! \, k_2! \, \ldots \, k_n! \, k_{n+1}!}{(k_1+k_2+\ldots+k_n+k_{n+1}+n)!} \; .$$

65. Evaluate the limit

$$L = \lim_{n \to \infty} \frac{1}{n} \sum_{k=1}^{n} \left(\left[\frac{2\sqrt{n}}{\sqrt{k}} \right] - 2 \left[\frac{\sqrt{n}}{\sqrt{k}} \right] \right) \; .$$

Solution: We show that $L = \dfrac{\pi^2}{3} - 3$. For any real number r we have

$$[2r] - 2[r] = \begin{cases} 0 \, , & \text{if } [2r] \text{ is even,} \\ 1 \, , & \text{if } [2r] \text{ is odd .} \end{cases}$$

Hence we obtain

$$A(n) = \sum_{k=1}^{n} \left(\left[\frac{2\sqrt{n}}{\sqrt{k}} \right] - 2 \left[\frac{\sqrt{n}}{\sqrt{k}} \right] \right)$$

$$= \sum_{\substack{k=1 \\ [2\sqrt{\frac{n}{k}}] \text{ odd}}}^{n} 1$$

$$= \sum_{\substack{s=1 \\ [2\sqrt{\frac{n}{k}}]=2s+1}}^{f(n)} \sum_{k=1}^{n} 1 \quad ,$$

where $f(n) = \left[\dfrac{[2\sqrt{n}] - 1}{2} \right]$. Next we see that $\left[2\sqrt{\dfrac{n}{k}} \right] = 2s + 1$ if and only if

$$\frac{4n}{(2s+2)^2} < k \le \frac{4n}{(2s+1)^2} \quad ,$$

and thus

$$\sum_{\substack{k=1 \\ [2\sqrt{\frac{n}{k}}]=2s+1}}^{n} 1 = \left[\frac{4n}{(2s+1)^2} \right] - \left[\frac{4n}{(2s+2)^2} \right]$$

$$= \frac{4n}{(2s+1)^2} - \frac{4n}{(2s+2)^2} + E_1 \quad ,$$

where $|E_1| < 1$. Hence we have

$$\frac{1}{n} A(n) = 4 \sum_{s=1}^{f(n)} \left(\frac{1}{(2s+1)^2} - \frac{1}{(2s+2)^2} \right) + E_2 \quad ,$$

where

$$|E_2| \le \frac{f(n)}{n} \cdot |E_1| < \frac{f(n)}{n} \le \frac{\sqrt{n}}{n} = \frac{1}{\sqrt{n}} \quad .$$

Letting $n \to \infty$ gives

$$L = 4 \sum_{s=1}^{\infty} \left(\frac{1}{(2s+1)^2} - \frac{1}{(2s+2)^2} \right) \;,$$

that is,

$$L = 4 \left(\left(\frac{\pi^2}{8} - 1 \right) - \left(\frac{\pi^2}{24} - \frac{1}{4} \right) \right) = \frac{\pi^2}{3} - 3 \;,$$

as $\;\; \dfrac{1}{1^2} + \dfrac{1}{3^2} + \dfrac{1}{5^2} + \ldots = \dfrac{\pi^2}{8} \;$ and $\; \dfrac{1}{2^2} + \dfrac{1}{4^2} + \dfrac{1}{6^2} + \ldots = \dfrac{\pi^2}{24} \;.$

66. Let p and q be distinct primes. Let S be the sequence consisting of the members of the set

$$\{p^m q^n : m,n = 0,1,2,\ldots \}$$

arranged in increasing order. For any pair (a,b) of non-negative integers, give an explicit expression involving a, b, p and q for the position of $p^a q^b$ in the sequence S .

Solution: Without loss of generality we may suppose that $p < q$.
    Clearly $p^a q^b$ is the nth term of the sequence, where n is the number of pairs of integers (r,s) such that

$$p^r q^s \leq p^a q^b \;, \quad r \geq 0 \;, \quad s \geq 0 \;.$$

Set

$$k = a + \left[\frac{b \, \ln q}{\ln p} \right] \;,$$

so that p^k is the largest power of p less than or equal to $p^a q^b$. Then we have

$$n = \sum_{\substack{r,s=0 \\ p^r q^s \leq p^a q^b}}^{k} 1$$

$$= \sum_{\substack{r,s = 0 \\ r\,\ell n\,p + s\,\ell n\,q \,\leq\, a\,\ell n\,p + b\,\ell n\,q}}^{k} 1$$

$$= \sum_{r=0}^{k} \left(\left[\frac{a\,\ell n\,p + b\,\ell n\,q - r\,\ell n\,p}{\ell n\,q} \right] + 1 \right)$$

$$= \sum_{r=0}^{k} \left(b + \left[\frac{(a-r)\,\ell n\,p}{\ell n\,q} \right] + 1 \right) ,$$

so that the position of $p^a q^b$ is

$$(k+1)(b+1) + \sum_{r=0}^{k} \left[\frac{(a-r)\,\ell n\,p}{\ell n\,q} \right] .$$

67. Let p denote an odd prime and let Z_p denote the finite field consisting of the p elements $0,1,2,\ldots,p-1$. For a an element of Z_p , determine the number $N(a)$ of 2×2 matrices X , with entries from Z_p , such that

(67.0) $$X^2 = A , \quad \text{where} \quad A = \begin{bmatrix} a & 0 \\ 0 & a \end{bmatrix}.$$

Solution: It is convenient to introduce the notation

$$k(a) = \begin{cases} 1 , & \text{if } a = b^2 \text{ for some non-zero element } b \text{ of } Z_p , \\ 0 , & \text{if } a = 0 , \\ -1, & \text{otherwise .} \end{cases}$$

We will show that

(67.1) $$N(a) = \begin{cases} p^2 + p + 2 , & \text{if } k(a) = 1 , \\ p^2 , & \text{if } k(a) = 0 , \\ p^2 - p , & \text{if } k(a) = -1 . \end{cases}$$

We note that if $k(a) = 1$, so that there is a non-zero element b of Z_p such that $a = b^2$, then the only other solution of $a = x^2$ is $x = -b$.

Let $X = \begin{bmatrix} x & y \\ z & w \end{bmatrix}$, where x,y,z and w are elements of Z_p , be a matrix such that $X^2 = A$. Then we have

(67.2) $x^2 + yz = yz + w^2 = a$,

(67.3) $(x + w)y = (x + w)z = 0$.

We treat two cases according as (i) $x + w = 0$ or (ii) $x + w \neq 0$.

In case (i) the equations (67.2) and (67.3) become

(67.4) $x^2 + yz = a$.

If $k(a) = 1$, say $a = b^2$, $b \neq 0$, then all the solutions of (67.4) are given by

$$(x,y,z) = (\pm b,0,0) \ , \ (\pm b,t,0) \ , \ (\pm b,0,t) \ , \ (u,t,t^{-1}(b^2-u^2)) \ ,$$

where t denotes a non-zero element of Z_p and u denotes an element of Z_p not equal to $\pm b$. Thus there are $2 + 2(p-1) + 2(p-1) + (p-2)(p-1) = p^2 + p$ solutions (x,y,z,w) in this case.

If $k(a) = 0$, so $a = 0$, then all the solutions of (67.4) are given by

$$(x,y,z) = (0,0,0) \ , \ (0,t,0) \ , \ (0,0,t) \ , \ (t,u,-t^2u^{-1}) \ ,$$

where t and u denote non-zero elements of Z_p . Thus there are $1 + (p-1) + (p-1) + (p-1)^2 = p^2$ solutions (x,y,z,w) in this case.

If $k(a) = -1$, so that a is not a square in Z_p , then all solutions of (67.4) are given by

$$(x,y,z) = (0,t,at^{-1}) \ , \ (t,u,(a-t^2)u^{-1}) \ ,$$

where t and u are non-zero elements of Z_p . Thus there are $(p-1) + (p-1)^2 = p^2 - p$ solutions (x,y,z,w) in this case.

In case (ii) the equations (67.2) and (67.3) become

$$x = w \neq 0 \;,\; x^2 = a \;,\; y = z = 0 \;,$$

which clearly has two solutions if $k(a) = 1$, and no solutions if $k(a) = 0$ or -1 .

Hence the total number of solutions is given by (67.1).

68. Let n be a non-negative integer and let $f(x)$ be the unique differentiable function defined for all real x by

$$(68.0) \qquad\qquad (f(x))^{2n+1} + f(x) - x = 0 \;.$$

Evaluate the integral

$$\int_0^x f(t)\, dt \;.$$

for $x \geq 0$.

Solution: The function $y = f(x)$ defined by (68.0) passes through the origin and has a positive derivative for all x . Hence, there exists an inverse function $f^{-1}(x)$, defined for all x, and such that $f^{-1}(0) = 0$. Clearly we have $f^{-1}(x) = x^{2n+1} + x$. When $x > 0$, the graph of $f(x)$ lies in the first quadrant, as $f(0) = 0$ and f is increasing. Thus for all $x \geq 0$ we have

$$\int_0^x f(t)\, dt + \int_0^{f(x)} f^{-1}(t)\, dt = x\, f(x) \;.$$

Now

$$\int_0^{f(x)} f^{-1}(t)\, dt = \int_0^{f(x)} (t^{2n+1} + t)\, dt = \frac{(f(x))^{2n+2}}{2n+2} + \frac{(f(x))^2}{2} \;,$$

so that

$$\int_0^x f(t)\, dt = x\, f(x) - \frac{(f(x))^{2n+2}}{2n+2} - \frac{(f(x))^2}{2}$$

$$= \frac{2n+1}{2n+2}\, x\, f(x) - \frac{n}{2n+2}\, (f(x))^2 .$$

69. Let $f(n)$ denote the number of zeros in the usual decimal representation of the positive integer n , so that for example, $f(1009) = 2$. For $a > 0$ and N a positive integer, evaluate the limit

$$L = \lim_{N \to \infty} \frac{\ell n\ S(N)}{\ell n\ N} ,$$

where

$$S(N) = \sum_{k=1}^{N} a^{f(k)} .$$

Solution: Let ℓ be a non-negative integer. The integers between 10^ℓ and $10^{\ell+1}-1$ have $\ell+1$ digits of which the first is necessarily non-zero. The number of these integers with i $(0 \le i \le \ell)$ of their digits equal to zero is $\binom{\ell}{i} 9^{\ell-i+1}$.

Choose m to be the unique non-negative integer such that

$$10^m - 1 \le N < 10^{m+1} - 1 ,$$

so that

$$m = \left[\frac{\ell n\ (N+1)}{\ell n\ 10} \right] .$$

Then we have

$$S(10^m - 1) = \sum_{k=1}^{10^m - 1} a^{f(k)} = \sum_{i=0}^{m-1} \sum_{\substack{k=1 \\ f(k)=i}}^{10^m - 1} a^{f(k)}$$

$$= \sum_{\substack{i=0}}^{m-1} a^i \sum_{\substack{k=1 \\ f(k)=i}}^{10^m-1} 1$$

$$= \sum_{\substack{i=0}}^{m-1} a^i \sum_{\ell=0}^{m-1} \sum_{\substack{k=10^\ell \\ f(k)=i}}^{10^{\ell+1}-1} 1 \quad .$$

Appealing to the first remark we obtain

$$S(10^m-1) = \sum_{i=0}^{m-1} a^i \sum_{\ell=0}^{m-1} \binom{\ell}{i} 9^{\ell-i+1}$$

$$= \sum_{\ell=0}^{m-1} 9^{\ell+1} \sum_{i=0}^{\ell} \binom{\ell}{i} \left(\frac{a}{9}\right)^i$$

$$= \sum_{\ell=0}^{m-1} 9^{\ell+1} \left(1 + \frac{a}{9}\right)^\ell$$

$$= 9 \sum_{\ell=0}^{m-1} (a + 9)^\ell \quad ,$$

that is

$$S(10^m-1) = c((a+9)^m - 1) \quad ,$$

where $c = \dfrac{9}{a+8}$.

As

$$S(10^m-1) \leq S(N) < S(10^{m+1}-1) \quad ,$$

we obtain

$$c(a+9)^m \leq S(N)+c < c(a+9)^{m+1} \quad .$$

Taking logarithms and dividing by m , we get

$$\frac{\ell n\ c}{m} - \ell n(a+9) \leq \frac{1}{m}\ell n(S(N)+c) < \frac{\ell n\ c}{m} + \frac{(m+1)}{m}\ell n(a+9) \ .$$

Letting $N \to \infty$, so that $m \to \infty$, we deduce that

$$\lim_{N \to \infty} \frac{1}{m}\ell n(S(N)+c) = \ell n(a+9) \ .$$

Hence we have

$$\lim_{N \to \infty} \frac{1}{m}\ell n\ S(N)$$

$$= \lim_{N \to \infty} \frac{1}{m}\ell n(S(N)+c) - \frac{1}{m}\ell n\left(1 + \frac{c}{S(N)}\right)$$

$$= \ell n(a+9) \ ,$$

and so

$$\lim_{N \to \infty} \frac{\ell n\ S(N)}{\ell n\ N} = \frac{\ell n(a+9)}{\ell n\ 10} \ .$$

70. Let $n \geq 2$ be an integer and let k be an integer with $2 \leq k \leq n$. Evaluate

$$M = \max_{S} \left(\min_{1 \leq i \leq k-1} (a_{i+1}-a_i) \right) \ ,$$

where S runs over all selections $S = \{a_1, a_2, \ldots, a_k\}$ from $\{1, 2, \ldots, n\}$ such that $a_1 < a_2 < \ldots < a_k$.

Solution (due to J.F. Semple and L. Smith):

We show that $M = \left[\dfrac{n-1}{k-1}\right]$. We consider the selection S^* given by

$$a_i = (i-1)\left[\frac{n-1}{k-1}\right] + 1 , \quad 1 \leq i \leq k ,$$

which has

$$a_{i+1} - a_i = \left[\frac{n-1}{k-1}\right] , \quad 1 \leq i \leq k-1 .$$

Thus for S^* we have

$$\min_{1 \leq i \leq k-1} (a_{i+1} - a_i) = \left[\frac{n-1}{k-1}\right] ,$$

so that $M \geq \left[\dfrac{n-1}{k-1}\right]$. In order to prove equality, we suppose that there is a selection S with

$$\min_{1 \leq i \leq k-1} (a_{i+1} - a_i) \geq \left[\frac{n-1}{k-1}\right] + 1 .$$

Then we have

$$n-1 \geq a_k - a_1 = \sum_{i=1}^{k-1} (a_{i+1} - a_i) \geq (k-1)\left(\left[\frac{n-1}{k-1}\right] + 1\right) > n-1 ,$$

which is impossible. Hence any selection has $\min\limits_{1 \leq i \leq k-1} (a_{i+1} - a_i) < \left[\dfrac{n-1}{k-1}\right] + 1$, which proves the required result.

71. Let $az^2 + bz + c$ be a polynomial with complex coefficients such that a and b are non-zero. Prove that the zeros of this polynomial lie in the region

(71.0) $$|z| \leq \left|\frac{b}{a}\right| + \left|\frac{c}{b}\right| .$$

Solution: Note that

$$\left| \sqrt{b^2 - 4ac} \right| = |b| \left| \sqrt{1 - \frac{4ac}{b^2}} \right|$$

$$\leq |b| \sqrt{1 + \left| \frac{4ac}{b^2} \right|}$$

$$\leq |b| \left(1 + \left| \frac{2ac}{b^2} \right| \right)$$

$$= |b| + \left| \frac{2ac}{b} \right| ,$$

so that

$$\left| - \frac{b}{2a} \pm \frac{\sqrt{b^2 - 4ac}}{2a} \right| \leq \left| \frac{b}{2a} \right| + \left| \frac{b}{2a} \right| + \left| \frac{c}{b} \right| = \left| \frac{b}{a} \right| + \left| \frac{c}{b} \right| ,$$

and hence the solutions of $az^2 + bz + c = 0$ satisfy (71.0).

Second solution (due to L. Smith): Let $w \ (\neq -1)$ be a complex number.
The inequality

(71.1) $$|w+1| + \frac{|w|}{|w+1|} \geq 1 ,$$

is easily established, for if $|w+1| \geq 1$ then (71.1) clearly holds,
while if $0 < |w+1| < 1$ then

$$|w+1| + \frac{|w|}{|w+1|} > |w+1| + |w| > 1 .$$

Let z_1, z_2 be the roots of the given quadratic chosen so that
$|z_1| \leq |z_2|$. As a and b are non-zero, $z_2 \neq 0$ and $z_1 + z_2 \neq 0$,
and setting $w = z_1/z_2 \ (\neq -1)$ in (71.1) we obtain

$$|z_2| \leq |z_1 + z_2| + \frac{|z_1 z_2|}{|z_1 + z_2|} .$$

The inequality (71.0) follows as $z_1 + z_2 = - \frac{b}{a}$ and $z_1 z_2 = \frac{c}{a}$.

72. Determine a monic polynomial $f(x)$ with integral coefficients such that $f(x) \equiv 0 \pmod{p}$ is solvable for every prime p but $f(x) = 0$ is not solvable with x an integer.

Solution: If $p = 2$ or $p \equiv 1 \pmod 4$ the congruence $x^2 + 1 \equiv 0$ $\pmod p$ is solvable. If $p \equiv 3 \pmod 8$ the congruence $x^2 + 2 \equiv 0 \pmod p$ is solvable. If $p \equiv 7 \pmod 8$ the congruence $x^2 - 2 \equiv 0 \pmod p$ is solvable. Set

$$f(x) = (x^2+1)(x^2+2)(x^2-2) .$$

Clearly $f(x)$ is a monic polynomial with integral coefficients such that $f(x) = 0$ is not solvable with x an integer.

73. Let n be a fixed positive integer. Determine

$$M = \max_{\substack{0 \le x_k \le 1 \\ k=1,2,\ldots,n}} \sum_{1 \le i < j \le n} |x_i - x_j| .$$

Solution: Without loss of generality we may assume that

$$0 \le x_1 \le x_2 \le \ldots \le x_n \le 1 ,$$

so that

$$S = \sum_{1 \le i < j \le n} |x_i - x_j| = \sum_{1 \le i < j \le n} (x_j - x_i) .$$

The sum S has $n(n-1)/2$ terms. For each k, $1 \le k \le n$, x_k appears in $k-1$ terms in the left position and $n-k$ times in the right position. Hence we have

$$S = \sum_{k=1}^{n} x_k ((k-1) - (n-k)) = \sum_{k=1}^{n} x_k (2k-n-1) .$$

As $2k-n-1 < 0$, for $k < \dfrac{n+1}{2}$, we have

$$S \leq \sum_{\frac{n+1}{2} \leq k \leq n} x_k (2k-n-1) \leq \sum_{\frac{n+1}{2} \leq k \leq n} (2k-n-1) \ .$$

Thus, for n even, we have

$$S \leq \sum_{k=\frac{n}{2}+1}^{n} (2k-n-1) = 1+3+5+\ldots+(n-1) = n^2/4 \ ,$$

and for n odd

$$S \leq \sum_{k=\frac{n+1}{2}}^{n} (2k-n-1) = 2+4+6+\ldots+(n-1) = (n^2-1)/4 \ .$$

Thus we have

$$S \leq [n^2/4] \ .$$

For n even, the choice

$$x_1 = x_2 = \ldots = x_{\frac{n}{2}} = 0 \ , \quad x_{\frac{n}{2}+1} = \ldots = x_n = 1$$

gives

$$S = n^2/4$$

and for n odd, the choice

$$x_1 = x_2 = \ldots = x_{\frac{n-1}{2}} = 0 \ , \quad x_{\frac{n+1}{2}} = \ldots = x_n = 1$$

gives

$$S = (n^2-1)/4 \ .$$

This shows that

$$M = [n^2/4] \ .$$

74. Let $\{x_i : i = 1,2,\ldots,n\}$ and $\{y_i : i = 1,2,\ldots,n\}$ be two sequences of real numbers with

$$x_1 \geq x_2 \geq \ldots \geq x_n .$$

How must $y_1, \ldots, y_n$ be rearranged so that the sum

(74.0)
$$\sum_{i=1}^{n} (x_i - y_i)^2$$

is as small as possible?

Solution: Suppose the ith term y_i is smaller than the jth term y_j, $1 \leq i < j \leq n$. Interchanging y_i and y_j produces a new sequence

$$z_1, z_2, \ldots, z_n$$

with

$$z_k = \begin{cases} y_k & , \text{ if } k \neq i \text{ or } j , \\ y_j & , \text{ if } k = i , \\ y_i & , \text{ if } k = j . \end{cases}$$

Moreover we have

$$\sum_{k=1}^{n} (x_k - z_k)^2 = \sum_{k=1}^{n} (x_k - y_k)^2 + (x_i - y_j)^2 + (x_j - y_i)^2 - (x_i - y_i)^2 - (x_j - y_j)^2,$$

where

$$((x_i - y_j)^2 + (x_j - y_i)^2) - ((x_i - y_i)^2 + (x_j - y_j)^2)$$
$$= (y_i - y_j)(2x_i - y_i - y_j) + (y_j - y_i)(2x_j - y_i - y_j)$$
$$= 2(y_i - y_j)(x_i - x_j)$$
$$\leq 0 ,$$

so that

$$\sum_{k=1}^{n} (x_k - z_k)^2 \leq \sum_{k=1}^{n} (x_k - y_k)^2 .$$

Hence every such transposition decreases the size of the sum (74.0), and the smallest sum is obtained when the y_i are arranged in decreasing order.

75. Let p be an odd prime and let Z_p denote the finite field consisting of $0,1,2,\ldots,p-1$. Let g be a given function on Z_p with values in Z_p . Determine all functions f on Z_p with values in Z_p , which satisfy the functional equation

(75.0) $f(x) + f(x+1) = g(x)$

for all x in Z_p .

Solution: Replacing x by $x+k$ ($k = 0,1,2,\ldots,p-1$) in (75.0), we obtain

(75.1) $f(x+k) + f(x+k+1) = g(x+k)$.

Hence, using (75.1), we have

$$\sum_{k=0}^{p-1} (-1)^k g(x+k) = \sum_{k=0}^{p-1} (-1)^k (f(x+k) + f(x+k+1))$$

$$= \sum_{k=0}^{p-1} ((-1)^k f(x+k) - (-1)^{k+1} f(x+k+1))$$

$$= f(x) - (-1)^p f(x+p)$$

$$= 2f(x) ,$$

so that there is only one such function, namely,

$$f(x) = 2^{-1} \sum_{k=0}^{p-1} (-1)^k g(x+k) .$$

76. Evaluate the double integral

(76.0)
$$I = \int_0^1 \int_0^1 \frac{dxdy}{1 - xy} .$$

Solution: Set $S = \{(x,y) \mid 0 \leq x \leq 1, 0 \leq y \leq 1\}$.

For $0 < \varepsilon < 1$ set

$$R_1(\varepsilon) = \{(x,y) \mid 0 \leq x \leq 1-\varepsilon, 0 \leq y \leq 1\},$$

$$R_2(\varepsilon) = \{(x,y) \mid 1-\varepsilon \leq x \leq 1, 0 \leq y \leq 1-\varepsilon\},$$

$$R(\varepsilon) = R_1(\varepsilon) \cup R_2(\varepsilon),$$

so that

$$R(\varepsilon) = S - \{(x,y) \mid 1-\varepsilon \leq x \leq 1, 1-\varepsilon \leq y \leq 1\}.$$

Then, for $j = 1,2$, we set

$$I_j(\varepsilon) = \iint_{R_j(\varepsilon)} \frac{dxdy}{1 - xy} .$$

The function $\frac{1}{1 - xy}$ is continuous on the square S except for a discontinuity at the corner $(x,y) = (1,1)$, so that (76.0) becomes

$$I = \lim_{\varepsilon \to 0^+} \iint_{R(\varepsilon)} \frac{dxdy}{1 - xy} = \lim_{\varepsilon \to 0^+} (I_1(\varepsilon) + I_2(\varepsilon)) .$$

For n a positive integer and for $j = 1,2$, we set

$$J_j(\varepsilon,n) = \iint_{R_j(\varepsilon)} (1 + xy + x^2y^2 + \ldots + x^{n-1}y^{n-1}) \, dx \, dy ,$$

$$K_j(\varepsilon,n) = \iint_{R_j(\varepsilon)} \frac{x^n y^n}{1 - xy} \, dx \, dy .$$

As $1 + xy + x^2y^2 + \ldots + x^{n-1}y^{n-1} + \dfrac{x^n y^n}{1 - xy}$

$$= \frac{1 - x^n y^n}{1 - xy} + \frac{x^n y^n}{1 - xy}$$

$$= \frac{1}{1 - xy} \; ,$$

we have, for $j = 1,2$; $0 < \varepsilon < 1$; $n \geq 1$,

(76.1) $J_j(\varepsilon,n) + K_j(\varepsilon,n) = I_j(\varepsilon)$.

Next, as the largest value of xy on both $R_1(\varepsilon)$ and $R_2(\varepsilon)$ is $1-\varepsilon$, we have

$$\left| \frac{x^n y^n}{1 - xy} \right| \leq \frac{(1 - \varepsilon)^n}{\varepsilon} \; , \quad \text{on } R_1(\varepsilon) \text{ and } R_2(\varepsilon) \; ,$$

so that

$$|K_j(\varepsilon,n)| \leq \begin{cases} \dfrac{(1-\varepsilon)^n}{\varepsilon} \cdot (1-\varepsilon) \; , & j = 1 \; , \\[2ex] \dfrac{(1-\varepsilon)^n}{\varepsilon} \cdot \varepsilon(1-\varepsilon) \; , & j = 2 \; . \end{cases}$$

Hence for $0 < \varepsilon < 1$ we have

$$\lim_{n \to \infty} K_j(\varepsilon,n) = 0 \; , \quad j = 1, 2 \; .$$

Next, for $j = 1,2$, we have

$$J_j(\varepsilon,n) = \iint\limits_{R_j(\varepsilon)} \left(\sum_{m=0}^{n-1} x^m y^m \right) dx\, dy$$

$$= \sum_{m=0}^{n-1} \iint\limits_{R_j(\varepsilon)} x^m y^m \, dx\, dy \; ,$$

so that

$$J_1(\varepsilon,n) = \sum_{m=0}^{n-1} \int_0^{1-\varepsilon} x^m \, dx \cdot \int_0^1 y^m \, dy$$

$$= \sum_{m=0}^{n-1} \frac{(1-\varepsilon)^{m+1}}{m+1} \cdot \frac{1}{m+1}$$

$$= \sum_{m=0}^{n-1} \frac{(1-\varepsilon)^{m+1}}{(m+1)^2} \quad ,$$

giving

$$\lim_{n \to \infty} J_1(\varepsilon,n) = \sum_{m=0}^{\infty} \frac{(1-\varepsilon)^{m+1}}{(m+1)^2} \quad ,$$

and similarly

$$\lim_{n \to \infty} J_2(\varepsilon,n) = \sum_{m=0}^{\infty} \frac{(1-\varepsilon)^{m+1}}{(m+1)^2} - \sum_{m=0}^{\infty} \frac{(1-\varepsilon)^{2m+2}}{(m+1)^2} \quad .$$

Letting $n \to \infty$ in (76.1), we obtain

$$I_1(\varepsilon) = \sum_{m=0}^{\infty} \frac{(1-\varepsilon)^{m+1}}{(m+1)^2} \quad ,$$

$$I_2(\varepsilon) = \sum_{m=0}^{\infty} \frac{(1-\varepsilon)^{m+1}}{(m+1)^2} - \sum_{m=0}^{\infty} \frac{(1-\varepsilon)^{2m+2}}{(m+1)^2} \quad ,$$

so that

$$I_1(\varepsilon) + I_2(\varepsilon) = 2 \sum_{m=0}^{\infty} \frac{(1-\varepsilon)^{m+1}}{(m+1)^2} - \sum_{m=0}^{\infty} \frac{(1-\varepsilon)^{2m+2}}{(m+1)^2} \quad ,$$

Hence, by Abel's limit theorem, we have

$$\int_0^1 \int_0^1 \frac{dxdy}{1-xy} = \lim_{\varepsilon \to 0^+} (I_1(\varepsilon) + I_2(\varepsilon))$$

$$= 2 \sum_{m=0}^{\infty} \frac{1}{(m+1)^2} - \sum_{m=0}^{\infty} \frac{1}{(m+1)^2}$$

$$= \sum_{m=0}^{\infty} \frac{1}{(m+1)^2} \ ,$$

that is $I = \pi^2/6$.

77. Let a and b be integers and m an integer > 1 .
Evaluate

$$\left[\frac{b}{m}\right] + \left[\frac{a+b}{m}\right] + \left[\frac{2a+b}{m}\right] + \ldots + \left[\frac{(m-1)a+b}{m}\right] \ .$$

Solution: Our starting point is the identity

(77.1) $$\sum_{x=0}^{k-1} \left[\frac{x}{k} + e\right] = [ek] \ ,$$

where k is any positive integer and e is any real number. As
$\{y\} = y - [y]$, for any real y , (77.1) becomes

(77.2) $$\sum_{x=0}^{k-1} \left\{\frac{x}{k} + e\right\} = \frac{1}{2}(k-1) + \{ek\} \ .$$

For fixed k and e , $\left\{\frac{x}{k} + e\right\}$ is periodic in x with period k .
If c is chosen to be an integer such that $GCD(c,k) = 1$, the map-
ping $x \to cx$ is a bijection on a complete residue system modulo k .
Applying this bijection to (77.2), we obtain

(77.3) $$\sum_{x=0}^{k-1} \left\{\frac{cx}{k} + e\right\} = \frac{1}{2}(k-1) + \{ek\} \ .$$

We now choose $c = a/GCD(a,m)$ and $k = m/GCD(a,m)$, so that
$GCD(c,k) = 1$, and $e = b/m$. Then (77.3) becomes (keeping k in
place of $m/GCD(a,m)$ where convenient)

$$\sum_{x=0}^{k-1} \left\{ \frac{ax+b}{m} \right\} = \frac{1}{2} \left(\frac{m}{GCD(a,m)} - 1 \right) + \left\{ \frac{b}{GCD(a,m)} \right\} ,$$

and so

$$\sum_{x=0}^{m-1} \left\{ \frac{ax+b}{m} \right\} = \sum_{z=0}^{GCD(a,m)-1} \sum_{y=0}^{k-1} \left\{ \frac{a(y+kz) + b}{m} \right\}$$

$$= \sum_{z=0}^{GCD(a,m)-1} \sum_{y=0}^{k-1} \left\{ \frac{ay+b}{m} \right\}$$

$$= \frac{1}{2} (m - GCD(a,m)) + GCD(a,m) \left\{ \frac{b}{GCD(a,m)} \right\} .$$

Finally, we have

$$\sum_{x=0}^{m-1} \left[\frac{ax+b}{m} \right] = \frac{a}{2}(m-1) + b - \frac{1}{2}(m-GCD(a,m)) - GCD(a,m) \left\{ \frac{b}{GCD(a,m)} \right\} ,$$

that is

$$\sum_{x=0}^{m-1} \left[\frac{ax+b}{m} \right] = \frac{1}{2}(am - a - m + GCD(a,m)) + GCD(a,m) \left\{ \frac{b}{GCD(a,m)} \right\} .$$

Remarks: The identity (77.1) is given as a problem (with hints) on
           page 40 in <u>Number Theory</u> by J. Hunter, Oliver and Boyd, 1964.

78.   Let $a_1, \ldots, a_n$ be   n   (>1)   distinct real numbers.   Set

$$S = a_1^2 + \ldots + a_n^2 , \quad M = \min_{1 \le i < j \le n} (a_i - a_j)^2 .$$

Prove that

$$\frac{S}{M} \ge \frac{n(n-1)(n+1)}{12} .$$

Solution: Relabeling the   a's , so that $a_1 < a_2 < \ldots < a_n$,
            preserves the values of   S   and   M .

Let $\min\limits_{1 \leq i \leq n} a_i^2 = a_j^2$, where j is a fixed subscript. Then, we have

$$a_i > 0 \ , \quad \text{for} \ i > j \ ,$$

$$a_i < 0 \ , \quad \text{for} \ i < j \ .$$

Next, $\min\limits_{1 \leq i \leq n-1} (a_{i+1} - a_i) = \sqrt{\min\limits_{1 \leq i \leq n-1} (a_{i+1} - a_i)^2} = \sqrt{\min\limits_{1 \leq i < j \leq n} (a_i - a_j)^2}$

$= \sqrt{M}$.

Define $b_i = a_j + \sqrt{M} \ (i - j)$, $i = 1, 2, \ldots, n$, so that

$b_i = b_1 + \sqrt{M} \ (i - 1)$.

Then, for $i > j$, we have

$a_i = (a_i - a_{i-1}) + (a_{i-1} - a_{i-2}) + \ldots + (a_{j+1} - a_j) + a_j$

$\geq \sqrt{M} \ (i - j) + a_j$

$= b_i$

$\geq a_j \geq - a_i$,

that is $a_i \geq b_i \geq - a_i$, $i > j$.

Similarly we have

$$a_i \leq b_i \leq - a_i \ , \quad i < j \ .$$

Thus, we obtain $a_i^2 \geq b_i^2$ $(i = 1, 2, \ldots, n)$, and so

$$S = \sum_{i=1}^{n} a_i^2 \geq \sum_{i=1}^{n} b_i^2 = \sum_{i=1}^{n} (b_1 + \sqrt{M} \ (i - 1))^2$$

$$= n\left(b_1 + \sqrt{M} \, \frac{(n-1)}{2}\right)^2 + \frac{(n-1)n(n+1)}{12} \, M$$

$$\geq \frac{(n-1)n(n+1)}{12} \, M \ .$$

79. Let $x_1, \ldots, x_n$ be n real numbers such that

$$\sum_{k=1}^{n} |x_k| = 1 \quad , \quad \sum_{k=1}^{n} x_k = 0 .$$

Prove that

(79.0)
$$\left| \sum_{k=1}^{n} \frac{x_k}{k} \right| \leq \frac{1}{2} - \frac{1}{2n} .$$

Solution: For $1 \leq k \leq \frac{2n}{n+1}$ we have

$$0 \leq \frac{2}{k} - 1 - \frac{1}{n} \leq 1 - \frac{1}{n} ,$$

and for $\frac{2n}{n+1} \leq k \leq n$ we have

$$0 \leq 1 + \frac{1}{n} - \frac{2}{k} \leq 1 - \frac{1}{n} ,$$

so that

(79.1)
$$\left| \frac{2}{k} - 1 - \frac{1}{n} \right| \leq 1 - \frac{1}{n} , \quad 1 \leq k \leq n .$$

Thus, as $\sum_{k=1}^{n} x_k = 0$, we have

$$\left| \sum_{k=1}^{n} \frac{x_k}{k} \right| = \frac{1}{2} \left| \sum_{k=1}^{n} \left(\frac{2}{k} - 1 - \frac{1}{n} \right) x_k \right|$$

$$\leq \frac{1}{2} \sum_{k=1}^{n} \left| \frac{2}{k} - 1 - \frac{1}{n} \right| |x_k|$$

$$\leq \frac{1}{2} \left(1 - \frac{1}{n} \right) \sum_{k=1}^{n} |x_k| , \quad \text{by (79.1)} .$$

The inequality (79.0) now follows as $\sum_{k=1}^{n} |x_k| = 1$.

80. Prove that the sum of two consecutive odd primes is the product of at least three (possibly repeated) prime factors.

Solution: Let p_n denote the nth prime number, so that $p_1 = 2$, $p_2 = 3$, $p_3 = 5$, ... For $n \geq 2$, we consider $q_n = p_n + p_{n+1}$. Clearly q_n is even. If q_n has exactly one prime factor then $q_n = 2^k$ for some positive integer k, and as $q_n \geq 3 + 5 = 2^3$ we have $k \geq 3$ proving the result in this case. If q_n has exactly two distinct prime factors, then we have

$q_n = 2^k p^\ell$ for positive integers k, ℓ and an odd prime p. If $k \geq 2$ or $\ell \geq 2$ the result holds. If $k = \ell = 1$ then, as $p_{n+1} > p_n$, we have

$$p_n < \frac{1}{2}(p_n + p_{n+1}) = p < p_{n+1} ,$$

which is impossible as p_n and p_{n+1} are consecutive primes. This completes the proof.

81. Let $f(x)$ be an integrable function on the closed interval $[\pi/2, \pi]$ and suppose that

(81.0) $$\int_{\pi/2}^{\pi} f(x) \sin kx \, dx = \begin{cases} 0 , & 1 \leq k \leq n-1 , \\ 1 , & k = n . \end{cases}$$

Prove that $|f(x)| \geq \frac{1}{\pi \ln 2}$ on a set of positive measure.

Solution: From (81.0) we have

(81.1) $$\sum_{k=1}^{n} \int_{\pi/2}^{\pi} f(x) \sin kx \, dx = 1 .$$

Interchanging the order of summation and integration, and using the identity

$$\sum_{k=1}^{n} \sin kx = \frac{\cos \frac{x}{2} - \cos(n+\frac{1}{2})x}{2 \sin \frac{x}{2}} \quad , \quad 0 < x < 2\pi \quad ,$$

(81.1) becomes

$$\int_{\pi/2}^{\pi} f(x) \frac{(\cos \frac{x}{2} - \cos(n+\frac{1}{2}))}{2 \sin \frac{x}{2}} dx = 1 \quad .$$

Suppose $|f(x)| < \frac{1}{\pi \ln 2}$ on $[\frac{\pi}{2}, \pi]$, except for a set of measure 0. Then we have

$$1 \leq \int_{\pi/2}^{\pi} |f(x)| \frac{\left|\cos \frac{x}{2} - \cos(n+\frac{1}{2})x\right|}{2 \sin \frac{x}{2}} dx \quad ,$$

giving

(81.2) $$1 < \frac{1}{\pi \ln 2} \int_{\pi/2}^{\pi} \frac{dx}{\sin \frac{x}{2}} \quad .$$

Now Jordan's inequality implies that

$$\frac{x}{\pi} \leq \sin \frac{x}{2} \quad (0 \leq x \leq \pi) \quad .$$

and so, on $[\frac{\pi}{2}, \pi]$, we have

(81.3) $$\frac{1}{\sin \frac{x}{2}} \leq \frac{\pi}{x} \quad .$$

Using (81.3) in (81.2) we obtain

$$1 < \frac{1}{\ln 2} \int_{\pi/2}^{\pi} \frac{dx}{x} = 1 \quad ,$$

which is impossible. Hence $|f(x)| \geq \frac{1}{\pi \ln 2}$ on a set of positive measure.

82. For $n = 0,1,2,\ldots$, let

(82.0) $\qquad s_n = \sqrt[3]{a_n + \sqrt[3]{a_{n-1} + \sqrt[3]{a_{n-2} + \ldots + \sqrt[3]{a_0}}}}$

where $a_n = \frac{6n+1}{n+1}$. Show that $\lim_{n \to \infty} s_n$ exists and determine its value.

Solution: First we show by mathematical induction that the sequence $\{s_n : n = 0,1,2,\ldots\}$ is non-decreasing. Note that $s_0 = 1 < \sqrt[3]{\frac{9}{2}} = s_1$. Assume that $s_{n-1} \leq s_n$. Then we have

$$s_n = \sqrt[3]{a_n + s_{n-1}} \leq \sqrt[3]{a_{n+1} + s_n} = s_{n+1}.$$

Next we show, also by induction, that the sequence $\{s_n : n = 0,1,2,\ldots\}$ is bounded above by 2. Clearly $s_0 = 1 < 2$. Assume that $s_{n-1} < 2$. Then we have

$$s_n = \sqrt[3]{a_n + s_{n-1}} < \sqrt[3]{6 + 2} = 2.$$

Thus $L = \lim_{n \to \infty} s_n$ exists. Letting $n \to \infty$ in $s_n^3 = a_n + s_{n-1}$ we obtain $L^3 = 6 + L$, so that $L = 2$.

83. Let $f(x)$ be a non-negative strictly increasing function on the interval $[a,b]$, where $a < b$. Let $A(x)$ denote the area below the curve $y = f(x)$ and above the interval $[a,x]$, where $a \leq x \leq b$, so that $A(a) = 0$.

Let $F(x)$ be a function such that $F(a) = 0$ and

(83.0) $\qquad (x' - x)f(x) < F(x') - F(x) < (x' - x)f(x')$

for all $a \leq x < x' \leq b$. Prove that $A(x) = F(x)$ for $a \leq x \leq b$.

Solution: Clearly $A(x)$ satisfies the inequality (83.0). Assume that $A(x)$ and $F(x)$ are not identical on $[a,b]$. Then there exists c with $a < c \leq b$ such that $A(c) \neq F(c)$. We partition the interval $[a,c]$ by

$$x_k = a + \frac{k(c-a)}{n} ,$$

where n is a positive integer and $k = 0,1,2,\ldots,n$. Then we have

$$\frac{(c-a)}{n} f(x_{k-1}) < A(x_k) - A(x_{k-1}) < \frac{(c-a)}{n} f(x_k) \quad (k = 1,2,\ldots,n) .$$

Summing from $k = 1$ to $k = n$, we obtain

(83.1)
$$\frac{c-a}{n} \sum_{k=0}^{n-1} f(x_k) < A(c) < \frac{c-a}{n} \sum_{k=1}^{n} f(x_k) .$$

Similarly we have

(83.2)
$$\frac{c-a}{n} \sum_{k=0}^{n-1} f(x_k) < F(c) < \frac{c-a}{n} \sum_{k=1}^{n} f(x_k) .$$

From (83.1) and (83.2) we obtain

$$\left| A(c) - F(c) \right| < ((c-a)/n)(f(c) - f(a))$$

so that (as $A(c) \neq F(c)$)

$$n < \frac{(c-a)(f(c) - f(a))}{|A(c) - F(c)|}$$

This is a contradicton for sufficiently large positive integers n .

84. Let a and b be two given positive numbers with $a < b$. How should the number r be chosen in the interval $[a,b]$ in order to minimize

(84.0)
$$M(r) = \max_{a \leq x \leq b} \left| \frac{r - x}{x} \right| \quad ?$$

Solution: For $a \leq r \leq b$ we have

$$\left| \frac{r - x}{x} \right| = \begin{cases} \dfrac{r}{x} - 1 , & \text{if } a \leq x \leq r , \\[2mm] 1 - \dfrac{r}{x} , & \text{if } r \leq x \leq b , \end{cases}$$

and so

$$M(r) = \max \left(\frac{r}{a} - 1 , 1 - \frac{r}{b} \right) .$$

Thus, for any c in the interval $[a,b]$, we have

$$\min_{a \leq r \leq b} M(r) = \min_{a \leq r \leq b} \max \left(\frac{r}{a} - 1 , 1 - \frac{r}{b} \right)$$

$$= \min \left[\min_{a \leq r \leq c} \max \left(\frac{r}{a} - 1 , 1 - \frac{r}{b} \right) , \right.$$
$$\left. \min_{c \leq r \leq b} \max \left(\frac{r}{a} - 1 , 1 - \frac{r}{b} \right) \right] .$$

Choosing c to be the point in $[a,b]$ such that

$$\frac{c}{a} - 1 = 1 - \frac{c}{b} ,$$

that is

$$c = \frac{2ab}{a+b} ,$$

we have

$$\min_{a \leq r \leq c} \max \left(\frac{r}{a} - 1 , 1 - \frac{r}{b} \right) = \min_{a \leq r \leq c} \left(1 - \frac{r}{b} \right) = 1 - \frac{c}{b}$$

and

$$\min_{c \leq r \leq b} \max \left(\frac{r}{a} - 1 , 1 - \frac{r}{b} \right) = \min_{c \leq r \leq b} \left(\frac{r}{a} - 1 \right) = \frac{c}{a} - 1$$

so that

$$\min_{a \leq r \leq b} M(r) = 1 - \frac{c}{b} \left(= \frac{c}{a} - 1 \right) = \frac{b-a}{a+b} ,$$

and the required r is $(2ab)/(a+b)$ . .

85. Let $\{a_n : n = 1, 2, \ldots \}$ be a sequence of positive real numbers with $\lim_{n \to \infty} a_n = 0$ and satisfying the condition $a_n - a_{n+1} > a_{n+1} - a_{n+2} > 0$. For any $\varepsilon > 0$, let N be a positive integer such that $a_N \leq 2\varepsilon$. Prove that $L = \sum_{k=1}^{\infty} (-1)^{k+1} a_k$ satisfies the inequality

(85.0)
$$\left| L - \sum_{k=1}^{N} (-1)^{k+1} a_k \right| < \varepsilon .$$

Solution: For n a positive integer, we define $S_n = \sum_{k=1}^{n} (-1)^{k+1} a_k$. We have

$$L = S_n + (-1)^n \sum_{r=1}^{\infty} (a_{n+r} - a_{n+r+1})$$

and

$$L = S_{n-1} + (-1)^{n-1} \sum_{r=1}^{\infty} (a_{n+r+1} - a_{n+r}) .$$

As $a_{n+r-1} - a_{n+r} > a_{n+r} - a_{n+r+1}$, we have

(85.1)
$$|S_n - L| < |S_{n-1} - L| .$$

Since $a_n = |S_n - S_{n-1}|$ and L lies between S_{n-1} and S_n, we have

(85.2)
$$a_n = |S_n - L| + |S_{n-1} - L| .$$

Taking $n = N$, where $a_N < 2\varepsilon$, we obtain

$$|S_N - L| < \varepsilon$$

as required.

36. Determine all positive continuous functions $f(x)$ defined on the interval $[0,\pi]$ for which

(86.0) $\int_0^\pi f(x) \cos nx \, dx = (-1)^n (2n+1)$, $n = 0,1,2,3,4$.

Solution: We begin with the identities

$$\cos 2x = 2 \cos^2 x - 1 ,$$
$$\cos 3x = 4 \cos^3 x - 3 \cos x ,$$
$$\cos 4x = 8 \cos^4 x - 8 \cos^2 x + 1 .$$

Hence

$$\cos 4x + 4 \cos 3x + 16 \cos 2x + 28 \cos x + 23$$
$$= 8 \cos^4 x + 16 \cos^3 x + 24 \cos^2 x + 16 \cos x + 8$$
$$= 8(\cos^2 x + \cos x + 1)^2 ,$$

and so

$$8 \int_0^\pi f(x)(\cos^2 x + \cos x + 1)^2 \, dx$$
$$= 9 + 4(-7) + 16(5) + 28(-3) + 23(1)$$
$$= 0 ,$$

which is impossible as $f(x)$ is positive on $[0,\pi]$. Hence there are no positive functions $f(x)$ satisfying (86.0).

87. Let P and P' be points on opposite sides of a non-circular ellipse E such that the tangents to E through P and P' respectively are parallel and such that the tangents and normals to E at P and P' determine a rectangle R of maximum area. Determine the equation of E with respect to a rectangular coordinate system, with origin at the centre of E and whose y-axis is parallel to the longer side of R .

Solution: We choose initially a coordinate system such that the
equation of E is $\frac{x^2}{a^2} + \frac{y^2}{b^2} = 1$, $a > b > 0$. The points
$Q = (a \cos t , b \sin t)$ $(0 \leq t \leq 2\pi)$ and $Q' = (-a \cos t , -b \sin t)$
lie on E and the tangents to E through Q and Q' are parallel.

We treat the case $0 \leq t \leq \pi/2$ as the other cases $\pi/2 \leq t \leq \pi$,
$\pi \leq t \leq 3\pi/2$, $3\pi/2 \leq t \leq 2\pi$ can be handled by appropriate reflec-
tions.

Let the normals through Q and Q' meet the tangents through
Q' and Q at T and T' respectively. Our first aim is to choose
t so that the area of the rectangle QTQ'T' is maximum. The slope
of the tangent to E at Q' is $\frac{-b \cos t}{a \sin t}$, and so the equations of
the lines Q'T and QT are respectively $b \cos t\, x + a \sin t\, y + ab = 0$
and $a \sin t\, x - b \cos t\, y - (a^2 - b^2) \sin t \cos t = 0$. Thus the lengths
$|QT|$ and $|Q'T|$ are given by

$$|QT| = \frac{2ab}{\sqrt{a^2 \sin^2 t + b^2 \cos^2 t}} \quad , \quad |Q'T| = \frac{2(a^2 - b^2)\sin t \cos t}{\sqrt{a^2 \sin^2 t + b^2 \cos^2 t}} .$$

The area of the rectangle QTQ'T' is clearly

$$\frac{4ab(a^2 - b^2) \tan t}{a^2 \tan^2 t + b^2} ,$$

whose maximum value $2(a^2 - b^2)$ is attained when $\tan t = \frac{b}{a}$. In this
case

$$\frac{|QT|}{|Q'T|} = \frac{a^2 + b^2}{a^2 - b^2} > 1 ,$$

so that R is not a square. Thus P is the point $\left(\frac{a^2}{\sqrt{a^2 + b^2}} , \frac{b^2}{\sqrt{a^2 + b^2}} \right)$
and the slope of the tangent at P is -1 . Rotating the axes
through $\pi/4$ clockwise by means of the orthogonal transformation
$(x,y) \to (X,Y)$, where $X = \frac{1}{\sqrt{2}} (x-y)$, $Y = \frac{1}{\sqrt{2}} (x+y)$, we find the equation of the required ellipse is

$$\frac{(X+Y)^2}{2a^2} + \frac{(X-Y)^2}{2b^2} = 1 .$$

88. If four distinct points lie in the plane such that any three of them can be covered by a disk of unit radius, prove that all four points may be covered by a disk of unit radius.

Solution: We first prove the following special case of Helly's theorem: If D_i (i = 1,2,3,4) are four disks in the plane such that any three have non-empty intersection then all four have non-empty intersection. Choose points W,X,Y,Z in $D_1 \cap D_2 \cap D_3$, $D_1 \cap D_2 \cap D_4$, $D_1 \cap D_3 \cap D_4$, $D_2 \cap D_3 \cap D_4$ respectively. We consider two cases according as one of the points W,X,Y,Z is in or on the (possibly degenerate) triangle formed by the other three points, or not.

In the first case suppose that Z is in or on triangle WXY. Then the line segments WX,WY,XY belong to $D_1 \cap D_2$, $D_1 \cap D_3$, $D_1 \cap D_4$ respectively, so that triangle WXY belongs to D_1, and thus Z belongs to D_1. Hence Z is a point of $D_1 \cap D_2 \cap D_3 \cap D_4$.

In the second case WXYZ is a quadrilateral whose diagonals intersect at a point C inside WXYZ. Without loss of generality we may suppose that C is the intersection of WY and XZ. Now the line segments WY and XZ belong to $D_1 \cap D_3$ and $D_2 \cap D_4$ respectively. Thus C is both in $D_1 \cap D_3$ and in $D_2 \cap D_4$, and so in $D_1 \cap D_2 \cap D_3 \cap D_4$.

To solve the problem let A,B,C,D be the four given distinct points. Let G be the centre of the unit disk to which A,B,C belong. Clearly the distances AG,BG,CG are all less than or equal to 1, and so G belongs in the three unit disks U_A, U_B, U_C centred at A,B,C respectively. Thus any three of the four disks U_A, U_B, U_C, U_D have a non-empty intersection, and so by the first result there is a point P in $U_A \cap U_B \cap U_C \cap U_D$. The unit disk centred at P contains A, B, C and D .

89. Evaluate the sum

$$S = \sum_{\substack{m=1 \\ m \neq n}}^{\infty} \sum_{n=1}^{\infty} \frac{1}{m^2 - n^2} .$$

Solution: For positive integers m and N with $N > m$, we have

$$A(m,N) = \sum_{\substack{n=1 \\ n \neq m}}^{N} \frac{1}{m^2-n^2} = -\frac{1}{2m} \sum_{\substack{n=1 \\ n \neq m}}^{N} \left(\frac{1}{n-m} - \frac{1}{n+m} \right)$$

$$= \frac{1}{2m} \left(S(N+m) - S(N-m) \right) - \frac{3}{4m^2} ,$$

where for $r = 1,2,3,\ldots$ we have set

$$S(r) = \sum_{k=1}^{r} \frac{1}{k} = \ell n\, r + c + E(r) ,$$

c denoting Euler's constant and the error term $E(r)$ satisfying

$$\left| E(r) \right| \leq \frac{A}{r} ,$$

for some absolute constant A . Then

$$\lim_{N \to \infty} (S(N+m) - S(N-m))$$

$$= \lim_{N \to \infty} \left[\ell n \left(\frac{N+m}{N-m} \right) + E(N+m) - E(N-m) \right]$$

$$= 0$$

and so

$$\lim_{N \to \infty} A(m,N) = -\frac{3}{4m^2} ,$$

and thus

$$S = \lim_{M \to \infty} \sum_{m=1}^{M} \lim_{N \to \infty} A(m,N)$$

$$= -\frac{3}{4} \lim_{M \to \infty} \sum_{m=1}^{M} \frac{1}{m^2}$$

$$= -\frac{\pi^2}{8} \ .$$

90. If n is a positive integer which can be expressed in the form $n = a^2 + b^2 + c^2$, where a,b,c are positive integers, prove that, for each positive integer k, n^{2k} can be expressed in the form $A^2 + B^2 + C^2$, where A,B,C are positive integers.

Solution: We begin by showing that if $m = x^2 + y^2 + z^2$, where x,y,z are positive integers, then $m^2 = X^2 + Y^2 + Z^2$, where X,Y,Z are positive integers. Without loss of generality we may choose $x \geq y \geq z$. Then the required X,Y,Z are given by

$$X = x^2 + y^2 - z^2 \ , \quad Y = 2xz \ , \quad Z = 2yz \ .$$

Letting $2k = 2^r(2s+1)$, where $r \geq 1$, $s \geq 0$, we have

$$n^{2s+1} = (n^s a)^2 + (n^s b)^2 + (n^s c)^2 \ ,$$

and applying the above argument successively we obtain

$$n^{2k} = (n^{2s+1})^{2^r} = X^2 + Y^2 + Z^2 \ ,$$

where X,Y,Z are positive integers.

91. Let G be the group generated by a and b subject to the relations $aba = b^3$ and $b^5 = 1$. Prove that G is abelian.

Solution: It suffices to show that a and b commute. The relation

$aba = b^3$ gives $b^{-1}ab = b^2a^{-1}$, and so

$$b^{-2}ab^2 = b^{-1}(b^{-1}ab)b$$
$$= b^{-1}(b^2a^{-1})b$$
$$= b^2(b^{-1}a^{-1}b)$$
$$= b^2(b^{-1}ab)^{-1}$$
$$= b^2(b^2a^{-1})^{-1}$$
$$= b^2ab^{-2} ,$$

giving

$$ab^4 = b^4a .$$

Hence, as $b^5 = 1$, we obtain $ab = b^5ab = b(b^4a)b = ba$.

92. Let $\{a_n : n = 1,2,3,\ldots \}$ be a sequence of real numbers

satisfying $0 < a_n < 1$ for all n and such that $\sum\limits_{n=1}^{\infty} a_n$ diverges

while $\sum\limits_{n=1}^{\infty} a_n^2$ converges. Let f(x) be a function defined on [0,1]

such that f"(x) exists and is bounded on [0,1]. If $\sum\limits_{n=1}^{\infty} f(a_n)$

converges, prove that $\sum\limits_{n=1}^{\infty} |f(a_n)|$ also converges.

Solution: Applying the extended mean value theorem to f on the
            interval $[0,a_n]$, there exists w_n such that $0 < w_n < a_n$

and

$$f(a_n) = f(0) + a_n f'(0) + \frac{a_n^2}{2} f''(w_n) .$$

If $\sum\limits_{n=1}^{\infty} f(a_n)$ converges, then we must have $\lim\limits_{n \to \infty} f(a_n) = 0$, and so by continuity $f(0) = 0$. Next, as $|f''(x)| \leq M$, $0 \leq x \leq 1$, we have

$$\sum_{n=1}^{\infty} \left| \frac{a_n^2 f''(w_n)}{2} \right| \leq \frac{M}{2} \sum_{n=1}^{\infty} a_n^2 \text{ , so that both } \sum_{n=1}^{\infty} \frac{a_n^2 f''(w_n)}{2} \text{ and } \sum_{n=1}^{\infty} \left| \frac{a_n^2 f''(w_n)}{2} \right|$$

converge. Hence $f'(0) \sum_{n=1}^{\infty} a_n = \sum_{n=1}^{\infty} f(a_n) - \sum_{n=1}^{\infty} \frac{a_n^2 f''(w_n)}{2}$ converges, and

so as $\sum_{n=1}^{\infty} a_n$ diverges, we must have $f'(0) = 0$. Thus

$$\sum_{n=1}^{\infty} |f(a_n)| = \sum_{n=1}^{\infty} \left| \frac{a_n^2 f''(w_n)}{2} \right| \text{ converges.}$$

93. Let a,b,c be real numbers such that the roots of the cubic equation

(93.0) $x^3 + ax^2 + bx + c = 0$

are all real. Prove that these roots are bounded above by $(2\sqrt{a^2-3b} - a)/3$.

Solution: Let p,q,r be the three real roots of (93.0) chosen so that $p \geq q \geq r$. Then, as $p^3 + ap^2 + bp + c = 0$, we have

$$x^3 + ax^2 + bx + c \equiv (x-p)(x^2 + (p+a)x + (p^2+ap+b)) .$$

The quadratic polynomial $x^2 + (p+a)x + (p^2+ap+b)$ has q and r as its two real roots, and hence its discriminant is non-negative, that is

(93.1) $(p+a)^2 - 4(p^2+ap+b) \geq 0$.

Solving (93.1) for p we obtain

$$p \leq (2\sqrt{a^2-3b} - a)/3 , \quad \text{which completes the proof.}$$

94. Let $Z_5 = \{0,1,2,3,4\}$ denote the finite field with 5 elements. Let a,b,c,d be elements of Z_5 with $a \neq 0$. Prove that the number N of distinct solutions in Z_5 of the cubic equation

$$f(x) = a + bx + cx^2 + dx^3 = 0$$

is given by $N = 4 - R$, where R denotes the rank of the matrix

$$A = \begin{bmatrix} a & b & c & d \\ b & c & d & a \\ c & d & a & b \\ d & a & b & c \end{bmatrix}.$$

Solution: Define B to be the Vandermonde matrix

$$B = \begin{bmatrix} 1 & 1 & 1^2 & 1^3 \\ 1 & 2 & 2^2 & 2^3 \\ 1 & 3 & 3^2 & 3^3 \\ 1 & 4 & 4^2 & 4^3 \end{bmatrix},$$

so that

$$BA = \begin{bmatrix} f(1) & 1^{-1}f(1) & 1^{-2}f(1) & 1^{-3}f(1) \\ f(2) & 2^{-1}f(2) & 2^{-2}f(2) & 2^{-3}f(2) \\ f(3) & 3^{-1}f(3) & 3^{-2}f(3) & 3^{-3}f(3) \\ f(4) & 4^{-1}f(4) & 4^{-2}f(4) & 4^{-3}f(4) \end{bmatrix}.$$

As $a \neq 0$, the matrix BA has N zero rows, so that rank $BA \leq 4{-}N$.

Let

$$[f(r_i) \quad r_i^{-1}f(r_i) \quad r_i^{-2}f(r_i) \quad r_i^{-3}f(r_i)] \quad (i = 1,\ldots,4{-}N)$$

be the 4-N non-zero rows of BA, where $1 \leq r_1 < \ldots < r_{4-N} \leq 4$.

Clearly

$$\begin{vmatrix} f(r_1) & r_1^{-1}f(r_1) & \cdots & r_1^{-(3-N)}f(r_1) \\ \cdot & \cdot & \cdots & \cdot \\ f(r_{4-N}) & r_{4-N}^{-1}f(r_{4-N}) & \cdots & r_{4-N}^{-(3-N)}f(r_{4-N}) \end{vmatrix}$$

$$= f(r_1) \cdots f(r_{4-N}) \begin{vmatrix} 1 & r_1^{-1} & \cdots & r_1^{-(3-N)} \\ \cdot & \cdot & \cdots & \cdot \\ 1 & r_{4-N}^{-1} & \cdots & r_{4-N}^{-(3-N)} \end{vmatrix}$$

$$\neq 0 ,$$

so that the rank of BA is exactly 4-N . Finally, since B is invertible, we have

$$R = \text{rank } A = \text{rank } BA = 4-N ,$$

that is, N = 4 - R .

  95. Prove that

(95.0)
$$S = \sum_{\substack{m,n=1 \\ (m,n)=1}}^{\infty} \frac{1}{(mn)^2}$$

is a rational number.

Solution: We notice that

$$\left(\sum_{r=1}^{\infty} \frac{1}{r^2} \right)^2 = \sum_{r,s=1}^{\infty} \frac{1}{(rs)^2} = \sum_{d=1}^{\infty} \sum_{\substack{r,s=1 \\ \text{GCD}(r,s)=d}}^{\infty} \frac{1}{(rs)^2} .$$

Setting r = dm , s = dn , so that GCD(m,n) = 1 , we obtain

$$\left(\frac{\pi^2}{6}\right)^2 = \sum_{d=1}^{\infty} \frac{1}{d^4} \sum_{\substack{m,n=1 \\ \text{GCD}(m,n)=1}}^{\infty} \frac{1}{(mn)^2} ,$$

that is

$$\frac{\pi^4}{36} = \frac{\pi^4}{90} S , \quad S = \frac{5}{2} .$$

96. Prove that there does not exist a rational function $f(x)$ with real coefficients such that

(96.0) $$f\left(\frac{x^2}{x+1}\right) = p(x) ,$$

where $p(x)$ is a non-constant polynomial with real coefficients.

Solution: Suppose there exists a rational function $f(x)$ and a non-constant polynomial $p(x)$ (both with real coefficients) such that (96.0) holds. As $f(x)$ is the quotient of two polynomials, there exist complex numbers $a \ (\neq 0)$, a_1, $\ldots$, a_r, b_1, $\ldots$, b_s with $a_i \neq b_j$ such that

(96.1) $$f(x) = \frac{a(x-a_1)\ldots(x-a_r)}{(x-b_1)\ldots(x-b_s)} .$$

Since $p(x)$ is a non-constant polynomial, $f(x)$ can neither be constant nor a polynomial, and so $s \geq 1$.

From (96.0) and (96.1), we obtain

(96.2) $$a(x+1)^{s-r} \frac{(x^2-a_1 x-a_1)\ldots(x^2-a_r x-a_r)}{(x^2-b_1 x-b_1)\ldots(x^2-b_s x-b_s)} = p(x) .$$

If $s-r < 0$, $x+1$ divides $x^2-a_i x-a_i$ for some i, $1 \leq i \leq r$, and so $(-1)^2 - a_i(-1) - a_i = 0$, that is, $1 + a_i - a_i = 0$, which is clearly impossible, and thus $s \geq r$.

Now let $x-c$ be a factor of $x^2 - b_1 x - b_1$, so that

(96.3) $$c^2 - b_1 c - b_1 = 0 .$$

Clearly $c \neq -1$. As the left side of (96.2) is a polynomial, we must have $r \geq 1$ and $x-c \mid x^2 - a_i x - a_i$, for some i, $1 \leq i \leq r$, that is

(96.4) $$c^2 - a_i c - a_i = 0 .$$

From (96.3) and (96.4) we obtain

$$a_i = \frac{c^2}{c+1} = b_1 ,$$

which is a contradiction. Hence no such rational function $f(x)$ exists.

97. For n a positive integer, set

$$S(n) = \sum_{k=0}^{n} \frac{1}{\binom{n}{k}} .$$

Prove that

$$S(n) = \frac{n+1}{2^{n+1}} \sum_{k=1}^{n+1} \frac{2^k}{k} .$$

Solution: For $n \geq 2$, we have

$$\frac{2^{n+1}}{n+1} S(n) - \frac{2^n}{n} S(n-1)$$

$$= \frac{2^{n+1}}{n+1} \sum_{k=0}^{n} \frac{1}{\binom{n}{k}} - \frac{2^n}{n} \sum_{k=0}^{n-1} \frac{1}{\binom{n-1}{k}}$$

$$= \frac{2^{n+1}}{n+1} + 2^n \sum_{k=0}^{n-1} \left(\frac{2}{(n+1)\binom{n}{k}} - \frac{1}{n\binom{n-1}{k}} \right)$$

$$= \frac{2^{n+1}}{n+1} + 2^n \sum_{k=0}^{n-1} \left(\frac{2k!(n-k)!}{(n+1)!} - \frac{k!(n-k-1)!}{n!} \right)$$

$$= \frac{2^{n+1}}{n+1} + \frac{2^n}{(n+1)!} \sum_{k=0}^{n-1} k!(n-k-1)!(n-2k-1)$$

$$= \frac{2^{n+1}}{n+1} + \frac{2^n}{(n+1)!} \sum_{k=0}^{n-1} \left(k!(n-k)! - (k+1)!(n-k-1)! \right)$$

$$= \frac{2^{n+1}}{n+1} + \frac{2^n}{(n+1)!}(0!n! - n!0!)$$

$$= \frac{2^{n+1}}{n+1} ,$$

and so

$$\frac{2^{n+1}}{n+1} S(n) - \frac{2^2}{2} S(1) = \sum_{k=3}^{n+1} \frac{2^k}{k} ,$$

which gives the required result as

$$\frac{2^2}{2} S(1) = \frac{2}{1} + \frac{2^2}{2} .$$

98. Let $u(x)$ be a non-trivial solution of the differential equation

$$u'' + pu = 0 ,$$

defined on the interval $I = [1,\infty)$, where $p = p(x)$ is continuous on I . Prove that u has only finitely many zeros in any interval $[a,b]$, $1 \le a < b$.

(A zero of $u(x)$ is a point z , $1 \le z < \infty$, with $u(z) = 0$).

Solution: Let S denote the set of zeros of $u(x)$ on the interval
[a,b], $1 \leq a < b$. We will assume that S is infinite
and derive a contradiction. The Bolzano-Weierstrass theorem implies
that S' has at least one accumulation point, say c, in [a,b]. Hence,
there exists either a decreasing or increasing sequence of zeros
$\{x_n: n = 1,2,3,\ldots\}$ converging to c. As u is continuous we
have $u(c) = 0$. Applying the mean value theorem to u on the
intervals with end-points x_n and x_{n+1} $(n = 1,2,3,\ldots)$, there
exists a sequence $\{y_n: n = 1,2,3,\ldots\}$ with y_n lying between
x_n and x_{n+1} and $u'(y_n) = 0$ for $n = 1,2,3,\ldots$. By the contin-
uity of u' we see that $u'(c) = 0$ since the y_n's converge to c.
   Now define
$$q = q(x) = u^2 + u'^2 , \quad c \leq x \leq b .$$
Then $q(c) = 0$ and
$$q'(x) = 2uu'(1-p)$$
so that

(98.1) $$q' \leq (1 + |p|)(u^2 + u'^2) \leq Kq ,$$

where $|p| \leq K-1$ on [c,b]. From (98.1) we deduce that
$$q(x) \leq q(c)e^{K(x-c)} , \quad c \leq x \leq b .$$
However $q(x) \geq 0$ so that $q(c) = 0$ implies that $q(x) \equiv 0$ on
[c,b], that is $u(x) \equiv 0$ on [c,b].
   The proof will be completed by showing that $u(x) \equiv 0$ on [a,c].
We set
$$v(x) = u(a+c-x)$$
and
$$r(x) = p(a+c-x)$$
for $a \leq x \leq c$. Then v is a solution of the differential equation

$$v'' + rv = 0$$

satisfying $v(a) = 0$, $v'(a) = 0$. By the above argument we deduce that $v(x) \equiv 0$ on $[a,c]$, and thus $u(x) \equiv 0$ on $[a,c]$.

This shows that $u(x) \equiv 0$ on $[a,b]$, for any $b > a$, and so $u(x) \equiv 0$ on $[1,\infty)$, contrary to assumption.

99. Let P_j $(j = 0,1,2,\ldots,n-1)$ be n (≥ 2) equally spaced points on a circle of unit radius. Evaluate the sum

$$S(n) = \sum_{0 \leq j < k \leq n-1} |P_j P_k|^2 \, ,$$

where $|PQ|$ denotes the distance between the points P and Q .

Solution: Without loss of generality we may take P_j $(j = 0,1,2,\ldots,n-1)$ to be the point $\exp(2\pi j i/n)$ on the unit circle $|z| = 1$ in the complex plane. Then, for $0 \leq j < k \leq n-1$, we have

$$|P_j P_k|^2 = |\exp(2\pi j i/n) - \exp(2\pi k i/n)|^2$$
$$= 2 - \exp(2\pi(k-j)i/n) - \exp(-2\pi(k-j)i/n)$$

and so

$$S(n) = \sum_{j=0}^{n-2} \sum_{k=j+1}^{n-1} (2 - \exp(2\pi(k-j)i/n) - \exp(-2\pi(k-j)i/n))$$
$$= 2 \sum_{j=0}^{n-2} (n-1-j) - A_n - \overline{A}_n \, ,$$

where

$$A_n = \sum_{j=0}^{n-2} \sum_{k=j+1}^{n-1} \exp(2\pi(k-j)i/n)$$

and $\overline{A}_n$ denotes the complex conjugate of A_n . Now

$$A_n = \sum_{j=0}^{n-2} \frac{\exp(2\pi i/n) - \exp(2\pi i(n-j)/n)}{1 - \exp(2\pi i/n)}$$

$$= \frac{(n-1)\exp(2\pi i/n)}{1 - \exp(2\pi i/n)} - \frac{1}{1 - \exp(2\pi i/n)} \sum_{j=0}^{n-2} \exp(-2\pi i j/n)$$

$$= \frac{(n-1)\exp(2\pi i/n)}{1 - \exp(2\pi i/n)} - \frac{1}{1 - \exp(-2\pi i/n)}$$

$$= \frac{n \exp(2\pi i/n)}{1 - \exp(2\pi i/n)} ,$$

and so

$$A_n + \overline{A}_n = -n .$$

Hence we obtain

$$S(n) = 2(n-1)^2 - 2 \frac{(n-2)(n-1)}{2} + n$$

giving

$$S(n) = n^2 .$$

100. Let M be a 3×3 matrix with entries chosen at random from the finite field $Z_2 = \{0,1\}$. What is the probability that M is invertible?

<u>Solution</u>: Let $M = (a_{ij})$, $1 \le i , j \le 3$, so that

$$D = \det M = a_{11}A_{11} + a_{12}A_{12} + a_{13}A_{13} ,$$

where

$A_{11} = a_{22}a_{33} - a_{23}a_{32}$, $A_{12} = a_{23}a_{31} - a_{21}a_{33}$, $A_{13} = a_{21}a_{32} - a_{22}a_{31}.$

If $(A_{11}, A_{12}, A_{13}) = (0,0,0)$, the number of corresponding triples (a_{11}, a_{12}, a_{13}) such that $D = 0$ is 8 . For each triple $(A_{11}, A_{12}, A_{13}) \ne (0,0,0)$, the number of corresponding triples (a_{11}, a_{12}, a_{13}) with $D = 0$ is 4 . Hence the number N of matrices

M with D = 0 is

(100.1) $N = 8n + 4(64-n) = 4n + 256$,

where n is the number of sextuples $(a_{21}, a_{22}, a_{23}, a_{31}, a_{32}, a_{33})$ with $A_{11} = A_{12} = A_{13} = 0$.

If $a_{21} = a_{22} = a_{23} = 0$ there are 8 triples (a_{31}, a_{32}, a_{33}) with $A_{11} = A_{12} = A_{13} = 0$. For each triple $(a_{21}, a_{22}, a_{23}) \neq (0,0,0)$, there are 2 triples (a_{31}, a_{32}, a_{33}) with $A_{11} = A_{12} = A_{13} = 0$. Hence we have

$$n = 1 \times 8 + 7 \times 2 = 22 ,$$

and so $N = 344$.

The required probability is

$$\frac{512 - 344}{512} = \frac{168}{512} \doteq 0.328 .$$

ABBREVIATIONS

ADM   Archiv der Mathematik

AI   Analytic Inequalities by D.S. Mitrinović,
Springer-Verlag (1970)

AMM   American Mathematical Monthly

BLMS   Bulletin of the London Mathematical Society

CF   Convex Figures by I.M. Yaglom and V.G.
Boltyanskii, Holt, Rinehart and Winston (1961)

CM   Crux Mathematicorum (formerly Eureka)

CMB   Canadian Mathematical Bulletin

CN   Course Notes for Mathematics 69.112,
Carleton University (1984)

ETN   Elementary Theory of Numbers by W. Sierpiński,
Warsaw (1964)

GCEA   Oxford and Cambridge Schools Examination Board,
General Certificate Examination, Scholarship Level,
Mathematics and Higher Mathematics

GCEB   Oxford and Cambridge Schools Examination Board,
General Certificate Examination, Scholarship Level,
Mathematics for Science

HCM   Oxford and Cambridge Schools Examination Board,
Higher Certificate Mathematics (Group III)

IMO   International Mathematical Olympiad

JUM   Journal of Undergraduate Mathematics

MM   Mathematics Magazine

NMT   Nordisk Matematisk Tidskrift

PMA   Principles of Mathematical Analysis by W. Rudin,
McGraw-Hill (1964)

PSM   Publicacions, Secció de Matemàtiques, Universitat
Autònoma de Barcelona

TN   Theory of Numbers by G.B. Mathews, Chelsea N.Y.
(1961)

WLP   William Lowell Putnam Mathematical Competition

REFERENCES

Problem

01: suggested by Problem 3, IMO (1970).

02: see Problem 382, CM 4 (1978), p.250.

03: see Problem 304, CM 4 (1978), p.11.

04: see Problem 138, CM 2 (1976), p.68.

05: suggested by Problem 161, CM 2 (1976), p.135.

06: see Problem 19, PMA, p.129.

08: suggested by Problem 162 CM 2 (1976), p.135.

09: suggested by Problem 207, CM 3 (1977), p.10.

10: suggested by Problem 5, GCEA (1954), Paper I.

11: see Problem 1, HCM (1947), Paper 5.

13: see Problem 3, IMO (1983).

15: see Problem 4, GCEA (1952), Paper V.

16: see Problem 1, GCEA (1952), Paper V.

17: see Problem 9, GCEA (1955), Paper V.

18: see Problem 4, GCEA (1955), Paper V.

19: suggested by Problem 2, HCM (1949), Paper 7.

20: based on Problem 6, GCEA (1952), Paper V.

21: see Problem 2(i), GCEB (1955), Paper V.

23: suggested by Problem 18, TN, p.318.

24: suggested by Problem J-2, CM 6 (1980), p.145.

26: suggested by Problem A-3, WLP (1980).

27: see Problem 3, HCM (1947), Paper 5.

28: suggested by Problem 2(ii), GCEB (1955), Paper V.

29: suggested by Problem 916, CM 10 (1984), p.54.

30: see AMM 76 (1969), pp.1130-1131.

31: see Problem 528, CM 7 (1981), p.90.

32: based on Problem 4, GCEA (1953), Paper VII.

33: see Problem 3, HCM (1948), Paper 7.

36: suggested by Problem 124, CM 2 (1976), p.119.

37: based on Problem 8, GCEA (1952), Paper V.

38: based on Problem 4, HCM (1943), Paper 5.

40: see Problem 10(ii), GCEA (1954), Paper VI.

41: based on Problem 5, HCM (1944), Paper 5.

42:   see Problem 853, CM 9 (1983), p.178.
43:   see Problem 3(ii), HCM (1949), Paper 5.
44:   see Problem 767, CM 9 (1983), p.281.
45:   see Problem 5, GCEA (1955), Paper V.
48:   suggested by Problem A-6, WLP (1973).
49:   based on Problem B-6, WLP (1974).
50:   suggested by Problem 4, HCM (1949), Paper 7.
53:   based on MM 44 (1971), pp.9-10.
54:   see Problem 602, CM 8 (1982), p.17.
55:   based on ETN, pp.124-125.
57:   see Problem 7, HCM (1948), Paper 5.
58:   suggested by Problem 6, CM 7 (1981), p.236.
59:   see Problem 5, HCM (1943), Paper 7.
61:   suggested by Problem B-2, WLP (1984).
63:   suggested by Problem A-3, WLP (1978).
64:   based on Problem A-5, WLP (1984).
65:   suggested by Problem B-1, WLP (1976).
66:   suggested by Problem 326, CM 4 (1978), p.66.
67:   suggested by Problem B-5, WLP (1968).
69:   suggested by Problem 3, CM 7 (1981), p.268.
71:   see AI, p.221.
72:   for a related result see AMM 76 (1969), p.1125.
73:   see Problem 4 (Afternoon Session), WLP (1961).
74:   suggested by Problem 1, IMO (1975).
75:   suggested by Problem B-2, WLP (1971).
76:   for more general results see BLMS 11 (1979),
      pp.268-272.
77:   based on ADM 25 (1974), pp.41-44.
78:   see Problem E2032, AMM 75 (1968), p.1124.
79:   see AI, pp.346-347.
80:   see Problem 21, CN, p.A72*3/4.
81:   suggested by Problem A-6, WLP (1972).
82:   for related results see JUM 15 (1983), pp.49-52.
83:   based on AMM 73 (1966), pp.477-483.
84:   based on AMM 57 (1950), pp.26-28.
85:   based on AMM 69 (1962), pp. 215-217.
86:   suggested by Problem 2 (Morning Session), WLP (1964).

88: see CF, Chapter 2.
90: based on PSM, 28 (1984), pp.75-80.
91: based on Problem 83, CMB 7 (1964), p.306.
92: suggested by Problem 1060, MM 52 (1979), p.46.
93: see Problem 105, CMB 9 (1966), p.532.
95: see Problem 171, CMB 15 (1972), p.313.
96: see Problem 178, CMB 15 (1972), pp.614-615.
97: see NMT 29 (1947), pp.97-103.
99: suggested by Problem 1104, MM 53 (1980), p.244.

A CATALOG OF SELECTED
DOVER BOOKS
IN SCIENCE AND MATHEMATICS

A CATALOG OF SELECTED
DOVER BOOKS
IN SCIENCE AND MATHEMATICS

QUALITATIVE THEORY OF DIFFERENTIAL EQUATIONS, V.V. Nemytskii and V.V. Stepanov. Classic graduate-level text by two prominent Soviet mathematicians covers classical differential equations as well as topological dynamics and ergodic theory. Bibliographies. 523pp. 5⅜ × 8½. 65954-2 Pa. $14.95

MATRICES AND LINEAR ALGEBRA, Hans Schneider and George Phillip Barker. Basic textbook covers theory of matrices and its applications to systems of linear equations and related topics such as determinants, eigenvalues and differential equations. Numerous exercises. 432pp. 5⅜ × 8½. 66014-1 Pa. $10.95

QUANTUM THEORY, David Bohm. This advanced undergraduate-level text presents the quantum theory in terms of qualitative and imaginative concepts, followed by specific applications worked out in mathematical detail. Preface. Index. 655pp. 5⅜ × 8½. 65969-0 Pa. $14.95

ATOMIC PHYSICS (8th edition), Max Born. Nobel laureate's lucid treatment of kinetic theory of gases, elementary particles, nuclear atom, wave-corpuscles, atomic structure and spectral lines, much more. Over 40 appendices, bibliography. 495pp. 5⅜ × 8½. 65984-4 Pa. $12.95

ELECTRONIC STRUCTURE AND THE PROPERTIES OF SOLIDS: The Physics of the Chemical Bond, Walter A. Harrison. Innovative text offers basic understanding of the electronic structure of covalent and ionic solids, simple metals, transition metals and their compounds. Problems. 1980 edition. 582pp. 6⅛ × 9¼. 66021-4 Pa. $16.95

BOUNDARY VALUE PROBLEMS OF HEAT CONDUCTION, M. Necati Özisik. Systematic, comprehensive treatment of modern mathematical methods of solving problems in heat conduction and diffusion. Numerous examples and problems. Selected references. Appendices. 505pp. 5⅜ × 8½. 65990-9 Pa. $12.95

A SHORT HISTORY OF CHEMISTRY (3rd edition), J.R. Partington. Classic exposition explores origins of chemistry, alchemy, early medical chemistry, nature of atmosphere, theory of valency, laws and structure of atomic theory, much more. 428pp. 5⅜ × 8½. (Available in U.S. only) 65977-1 Pa. $11.95

A HISTORY OF ASTRONOMY, A. Pannekoek. Well-balanced, carefully reasoned study covers such topics as Ptolemaic theory, work of Copernicus, Kepler, Newton, Eddington's work on stars, much more. Illustrated. References. 521pp. 5⅜ × 8½. 65994-1 Pa. $12.95

PRINCIPLES OF METEOROLOGICAL ANALYSIS, Walter J. Saucier. Highly respected, abundantly illustrated classic reviews atmospheric variables, hydrostatics, static stability, various analyses (scalar, cross-section, isobaric, isentropic, more). For intermediate meteorology students. 454pp. 6⅛ × 9¼. 65979-8 Pa. $14.95

ASYMPTOTIC METHODS IN ANALYSIS, N.G. de Bruijn. An inexpensive, comprehensive guide to asymptotic methods—the pioneering work that teaches by explaining worked examples in detail. Index. 224pp. 5⅜ × 8½. 64221-6 Pa. $7.95

OPTICAL RESONANCE AND TWO-LEVEL ATOMS, L. Allen and J.H. Eberly. Clear, comprehensive introduction to basic principles behind all quantum optical resonance phenomena. 53 illustrations. Preface. Index. 256pp. 5⅜ × 8½. 65533-4 Pa. $8.95

COMPLEX VARIABLES, Francis J. Flanigan. Unusual approach, delaying complex algebra till harmonic functions have been analyzed from real variable viewpoint. Includes problems with answers. 364pp. 5⅜ × 8½. 61388-7 Pa. $9.95

ATOMIC SPECTRA AND ATOMIC STRUCTURE, Gerhard Herzberg. One of best introductions; especially for specialist in other fields. Treatment is physical rather than mathematical. 80 illustrations. 257pp. 5⅜ × 8½. 60115-3 Pa. $6.95

APPLIED COMPLEX VARIABLES, John W. Dettman. Step-by-step coverage of fundamentals of analytic function theory—plus lucid exposition of five important applications: Potential Theory; Ordinary Differential Equations; Fourier Transforms; Laplace Transforms; Asymptotic Expansions. 66 figures. Exercises at chapter ends. 512pp. 5⅜ × 8½. 64670-X Pa. $12.95

ULTRASONIC ABSORPTION: An Introduction to the Theory of Sound Absorption and Dispersion in Gases, Liquids and Solids, A.B. Bhatia. Standard reference in the field provides a clear, systematically organized introductory review of fundamental concepts for advanced graduate students, research workers. Numerous diagrams. Bibliography. 440pp. 5⅜ × 8½. 64917-2 Pa. $11.95

UNBOUNDED LINEAR OPERATORS: Theory and Applications, Seymour Goldberg. Classic presents systematic treatment of the theory of unbounded linear operators in normed linear spaces with applications to differential equations. Bibliography. 199pp. 5⅜ × 8½. 64830-3 Pa. $7.95

LIGHT SCATTERING BY SMALL PARTICLES, H.C. van de Hulst. Comprehensive treatment including full range of useful approximation methods for researchers in chemistry, meteorology and astronomy. 44 illustrations. 470pp. 5⅜ × 8½. 64228-3 Pa. $11.95

CONFORMAL MAPPING ON RIEMANN SURFACES, Harvey Cohn. Lucid, insightful book presents ideal coverage of subject. 334 exercises make book perfect for self-study. 55 figures. 352pp. 5⅜ × 8¼. 64025-6 Pa. $11.95

OPTICKS, Sir Isaac Newton. Newton's own experiments with spectroscopy, colors, lenses, reflection, refraction, etc., in language the layman can follow. Foreword by Albert Einstein. 532pp. 5⅜ × 8½. 60205-2 Pa. $11.95

GENERALIZED INTEGRAL TRANSFORMATIONS, A.H. Zemanian. Graduate-level study of recent generalizations of the Laplace, Mellin, Hankel, K. Weierstrass, convolution and other simple transformations. Bibliography. 320pp. 5⅜ × 8½. 65375-7 Pa. $8.95

SPECIAL FUNCTIONS, N.N. Lebedev. Translated by Richard Silverman. Famous Russian work treating more important special functions, with applications to specific problems of physics and engineering. 38 figures. 308pp. 5⅜ × 8½.
60624-4 Pa. $9.95

OBSERVATIONAL ASTRONOMY FOR AMATEURS, J.B. Sidgwick. Mine of useful data for observation of sun, moon, planets, asteroids, aurorae, meteors, comets, variables, binaries, etc. 39 illustrations. 384pp. 5⅜ × 8¼. (Available in U.S. only)
24033-9 Pa. $8.95

INTEGRAL EQUATIONS, F.G. Tricomi. Authoritative, well-written treatment of extremely useful mathematical tool with wide applications. Volterra Equations, Fredholm Equations, much more. Advanced undergraduate to graduate level. Exercises. Bibliography. 238pp. 5⅜ × 8½.
64828-1 Pa. $8.95

POPULAR LECTURES ON MATHEMATICAL LOGIC, Hao Wang. Noted logician's lucid treatment of historical developments, set theory, model theory, recursion theory and constructivism, proof theory, more. 3 appendixes. Bibliography. 1981 edition. ix + 283pp. 5⅜ × 8½.
67632-3 Pa. $8.95

MODERN NONLINEAR EQUATIONS, Thomas L. Saaty. Emphasizes practical solution of problems; covers seven types of equations. ". . . a welcome contribution to the existing literature. . . ."—*Math Reviews*. 490pp. 5⅜ × 8½. 64232-1 Pa. $11.95

FUNDAMENTALS OF ASTRODYNAMICS, Roger Bate et al. Modern approach developed by U.S. Air Force Academy. Designed as a first course. Problems, exercises. Numerous illustrations. 455pp. 5⅜ × 8½.
60061-0 Pa. $9.95

INTRODUCTION TO LINEAR ALGEBRA AND DIFFERENTIAL EQUATIONS, John W. Dettman. Excellent text covers complex numbers, determinants, orthonormal bases, Laplace transforms, much more. Exercises with solutions. Undergraduate level. 416pp. 5⅜ × 8½.
65191-6 Pa. $10.95

INCOMPRESSIBLE AERODYNAMICS, edited by Bryan Thwaites. Covers theoretical and experimental treatment of the uniform flow of air and viscous fluids past two-dimensional aerofoils and three-dimensional wings; many other topics. 654pp. 5⅜ × 8½.
65465-6 Pa. $16.95

INTRODUCTION TO DIFFERENCE EQUATIONS, Samuel Goldberg. Exceptionally clear exposition of important discipline with applications to sociology, psychology, economics. Many illustrative examples; over 250 problems. 260pp. 5⅜ × 8½.
65084-7 Pa. $8.95

LAMINAR BOUNDARY LAYERS, edited by L. Rosenhead. Engineering classic covers steady boundary layers in two- and three-dimensional flow, unsteady boundary layers, stability, observational techniques, much more. 708pp. 5⅜ × 8½.
65646-2 Pa. $18.95

LECTURES ON CLASSICAL DIFFERENTIAL GEOMETRY, Second Edition, Dirk J. Struik. Excellent brief introduction covers curves, theory of surfaces, fundamental equations, geometry on a surface, conformal mapping, other topics. Problems. 240pp. 5⅜ × 8½.
65609-8 Pa. $8.95

CHALLENGING MATHEMATICAL PROBLEMS WITH ELEMENTARY SOLUTIONS, A.M. Yaglom and I.M. Yaglom. Over 170 challenging problems on probability theory, combinatorial analysis, points and lines, topology, convex polygons, many other topics. Solutions. Total of 445pp. 5⅜ × 8½. Two-vol. set.
Vol. I 65536-9 Pa. $7.95
Vol. II 65537-7 Pa. $7.95

FIFTY CHALLENGING PROBLEMS IN PROBABILITY WITH SOLUTIONS, Frederick Mosteller. Remarkable puzzlers, graded in difficulty, illustrate elementary and advanced aspects of probability. Detailed solutions. 88pp. 5⅜ × 8½.
65355-2 Pa. $4.95

EXPERIMENTS IN TOPOLOGY, Stephen Barr. Classic, lively explanation of one of the byways of mathematics. Klein bottles, Moebius strips, projective planes, map coloring, problem of the Koenigsberg bridges, much more, described with clarity and wit. 43 figures. 210pp. 5⅜ × 8½. 25933-1 Pa. $6.95

RELATIVITY IN ILLUSTRATIONS, Jacob T. Schwartz. Clear nontechnical treatment makes relativity more accessible than ever before. Over 60 drawings illustrate concepts more clearly than text alone. Only high school geometry needed. Bibliography. 128pp. 6⅛ × 9¼. 25965-X Pa. $7.95

AN INTRODUCTION TO ORDINARY DIFFERENTIAL EQUATIONS, Earl A. Coddington. A thorough and systematic first course in elementary differential equations for undergraduates in mathematics and science, with many exercises and problems (with answers). Index. 304pp. 5⅜ × 8½. 65942-9 Pa. $8.95

FOURIER SERIES AND ORTHOGONAL FUNCTIONS, Harry F. Davis. An incisive text combining theory and practical example to introduce Fourier series, orthogonal functions and applications of the Fourier method to boundary-value problems. 570 exercises. Answers and notes. 416pp. 5⅜ × 8½. 65973-9 Pa. $11.95

AN INTRODUCTION TO ALGEBRAIC STRUCTURES, Joseph Landin. Superb self-contained text covers "abstract algebra": sets and numbers, theory of groups, theory of rings, much more. Numerous well-chosen examples, exercises. 247pp. 5⅜ × 8½. 65940-2 Pa. $8.95

Prices subject to change without notice.
Available at your book dealer or write for free Mathematics and Science Catalog to Dept. GI, Dover Publications, Inc., 31 East 2nd St., Mineola, N.Y. 11501. Dover publishes more than 175 books each year on science, elementary and advanced mathematics, biology, music, art, literature, history, social sciences and other areas.